D1067602

METHODS OF MICROARRAY DATA ANALYSIS III

METHODS OF MICROARRAY DATA ANALYSIS III

Papers from CAMDA '02

edited by

Kimberly F. Johnson
Cancer Center Information Systems
Duke University Medical Center
Durham, NC

Simon M. Lin
Duke Bioinformatics Shared Resource
Duke University Medical Center
Durham, NC

KLUWER ACADEMIC PUBLISHERS
Boston / New York / Dordrecht / London

Distributors for North, Central and South America:
Kluwer Academic Publishers
101 Philip Drive
Assinippi Park
Norwell, Massachusetts 02061 USA
Telephone (781) 871-6600
Fax (781) 681-9045
E-Mail: kluwer@wkap.com

Distributors for all other countries:
Kluwer Academic Publishers Group
Post Office Box 322
3300 AH Dordrecht, THE NETHERLANDS
Telephone 31 786 576 000
Fax 31 786 576 254
E-Mail: services@wkap.nl

Electronic Services <http://www.wkap.nl>

Library of Congress Cataloging-in-Publication Data

A C.I.P. Catalogue record for this book is available
from the Library of Congress.

Title: METHODS OF MICROARRAY DATA ANALYSIS III
Editor: Kimberly F. Johnson and Simon M. Lin
ISBN: 1-4020-7582-0

Printed on acid-free paper.

Printed in the United States of America.

Contents

Contributing Authors

Abruzzo, Lynne V., University of Texas M.D. Anderson Cancer Center, Houston, TX
Al-Shahrour, Fátima, Centro Nacional de Investigaciones Oncológicas, (CNIO),
(Spanish National Cancer Centre), Madrid, Spain
Arbieva, Zarema, University of Illinois at Chicago, Chicago, IL
Arjas, Elja, Rolf Nevanlinna Institute, University of Helsinki, Finland
Bhattacharjee, Madhuchhanda, Rolf Nevanlinna Institute, University of Helsinki, Finland
Chu, Tzu-Ming, SAS Institute, Cary, NC
Churchill, Gary A., The Jackson Laboratory, Bar Harbor, Maine
Coombes, Kevin R., University of Texas M.D. Anderson Cancer Center, Houston, TX
Cui, Xiangqin, The Jackson Laboratory, Bar Harbor, Maine
Datta, D., Fox Chase Cancer Center, Philadelphia, PA
Deng, Shibing, SAS Institute, Cary, NC
Díaz-Uriarte, Ramón, Centro Nacional de Investigaciones Oncológicas, (CNIO), (Spanish National Cancer Centre), Madrid, Spain
Dopazo, Joaquín, Centro Nacional de Investigaciones Oncológicas, (CNIO), (Spanish National Cancer Centre), Madrid, Spain
Golemis, Erica A., Fox Chase Cancer Center, Philadelphia, PA
Gonye, Gregory E., Thomas Jefferson University, Philadelphia, PA
Hsieh, Wen-Ping, North Carolina State University, Raleigh, NC
Kachalo, Seman, University of Illinois at Chicago, Chicago, IL
Kossenkov, A. V., Fox Chase Cancer Center, Philadelphia, PA *and* Moscow Physical Engineering Institute, Moscow, Russian Federation
Liang, Jie, University of Illinois at Chicago, Chicago, IL
Liu, Ben, Bio-informatics Group Inc., Cary, NC
Moloshok, T. D., Fox Chase Cancer Center, Philadelphia, PA
Ochs, Michael F., Fox Chase Cancer Center, Philadelphia, PA
Pearson, Ronald K., Thomas Jefferson University, Philadelphia, PA

Pritchard, Colin, Fred Hutchinson Cancer Research Centre, Seattle, WA
Ptitsyn, Andrey, Pennington Biomedical Research Center
Rosner, Gary L., University of Texas M.D. Anderson Cancer Center, Houston, TX
Schwaber, James S, Thomas Jefferson University, Philadelphia, PA
Sillanpää, Mikko J., Rolf Nevanlinna Institute, University of Helsinki, Finland
Stivers, David N., University of Texas M.D. Anderson Cancer Center, Houston, TX
Wang, Jing, University of Texas M.D. Anderson Cancer Center, Houston, TX
Warren, Liling, Bio-informatics Group Inc., Cary, NC
Wolfinger, Russ, SAS Institute, Cary, NC
Xiao, Lianchun, The University of Texas MD Anderson Cancer Center, Houston, TX
Zhang, Li, The University of Texas MD Anderson Cancer Center, Houston, TX

Preface

As microarray technology has matured, data analysis methods have advanced as well. However, microarray results can vary widely from lab to lab as well as from chip to chip, with many opportunities for errors along the path from sample to data. The third CAMDA conference held in November of 2002 pointed out the increasing need for data quality assurance mechanisms through real world problems with the CAMDA datasets. Thus, the third volume of Methods of Microarray Data Analysis emphasizes many aspects of data quality assurance.

We highlight three tutorial papers to assist with a basic understanding of underlying principles in microarray data analysis, and add twelve papers presented at the conference. As editors, we have not comprehensively edited these papers, but have provided comments to the authors to encourage clarity and expansion of ideas. Each paper was peer-reviewed and returned to the author for further revision.

We do not propose these methods as the *de facto* standard for microarray analysis. But rather we present them as starting points for discussion to further the science of micrarray data analysis. The CAMDA conference continues to bring to light problems, solutions and new ideas to this arena and offers a forum for continued advancement of the art and science of microarray data analysis.

Kimberly Johnson

Simon Lin

Introduction

A comparative study of analytical methodologies using a standard data set has proven fruitful in microarray analysis. To provide a forum for these comparisons the third Critical Assessment of Microarray Data Analysis (CAMDA) conference was held in November, 2002. Over 170 researchers from eleven countries heard twelve presentations on topics such as data quality analysis, image analysis, data normalization, expression variance, cross hybridization and pattern searching. The conference has evolved in its third year, just as the science of microarrays has developed. While initial microarray data analysis techniques focused on classification exercises (CAMDA '00), and later on pattern extraction (CAMDA '01), this year's conference, by necessity, focused on data quality issues. This shift in focus follows the maturation of microarray technology as the detection of data quality problems has become a prerequisite for data analysis. Problems such as background noise determination, faulty fabrication processes, and, in our case, errors in data handling, were highlighted at the conference.

The CAMDA '02 conference provided a real-world lesson on data quality control and saw significant development of the cross-hybridization models. In this volume, we present three tutorial chapters and twelve paper presentations. First, Michael Ochs and Erica Golemis present a tutorial called "The Biology Behind Gene Expression." This discussion is for non-biologists who want to know more about an intelligent machine called the cell. This machinery is extremely complex and a glossary in this tutorial provides the novice with an overview of important terms related to microarrays while the rest of the paper details the biological processes that impact microarray analysis. Next is a tutorial on methods of data quality control by Kevin Coombes. We invited Dr. Coombes to submit this tutorial, titled "Monitoring the Quality of Microarray Experiments," as an expansion of his presentation at the conference, which is also prominently featured. The last tutorial is by Ronald Pearson titled "Outliers in Microarray Data Analysis." This tutorial addresses the issue of quality control by identifying outliers and suggesting methods to deal with technical and biological variations in microarray data.

As always, we are happy to highlight the paper voted by attendees as the Best Presentation. This year, the award went to:

David N. Stivers, Jing Wang, Gary L. Rosner, and Kevin R. Coombes
University of Texas M.D. Anderson Cancer Center, Houston, TX
"Organ-specific Differences in Gene Expression and UniGene Annotations Describing Source Material"

for their rigorous scrutiny of data quality before starting data analysis. Presented by Kevin Coombes, their paper not only revealed the existence of errors in the Project Normal data set, but also specified the exact nature of the problems and included the methods used to detect these problems. See below for more details on these data set errors.

CAMDA 2002 Data Sets

The scientific committee chose two data sets for CAMDA '02. The first, called Project Normal came from The Fred Hutchinson Cancer Center and it showed the variation of baseline gene expression levels in the liver, kidney and testis of six normal mice. By using a 5406-clone spotted cDNA microarray, Pritchard *et al.* concluded that replications are necessary in microarray experiments. The second data set came from the Latin Square Study at Affymetrix Inc. This benchmark data set was created to develop statistical algorithms for microarrays. Sets of fourteen genes with known concentrations were spiked into a complex background solution and hybridized on Affymetrix chips. Data was obtained with replicates and both Human and E. Coli chips were studied.

As mentioned above, there were errors in the Project Normal dataset that were undetected until CAMDA abstracts were submitted. Once we received the Stivers *et al.* abstract, we asked the original Project Normal authors to confirm their findings. The errors in the data set were verified and after much discussion among the Scientific Committee members, a decision was made to keep the contest going to allow the Stivers group to report and discuss their finding of data abnormalities at the conference. Actually, many groups revealed various aspects of the data abnormalities, but the Stivers group not only realized that a problem existed, but also identified the specific problem. Colin Pritchard, representing Project Normal, confirmed that indeed, the problems in the dataset were a result of incorrectly merging the data with the annotations, resulting in mismatched row/column combinations. In addition, several slides have a small number of misaligned grids. These problems affected about 1/3 of the genes (though different sets) in the testis and liver data. Pritchard also noted that a re-analysis of the data with the corrected data sets showed that the results were not notably different from the original conclusions. For the record, both the original and the corrected data sets are available at the CAMDA conference website for researchers who might be interested in "data forensics". We extend our thanks to Colin Pritchard, Li Hsu, and Peter Nelson at the Fred Hutchinson Cancer Center for their assistance and professionalism in handling this discovery and allowing the conference to proceed as planned. They were most gracious in their contributions to the conference.

Organization of this Volume

After presenting the three tutorial papers, naturally the first conference paper is the one voted as Best Presentation. We then divide the book into subject areas covering image analysis, data normalization, variance characterization, cross hybridization, and finally pattern searching. At the end of this introduction, you will also find a link to the web companion to this volume.

Analyzing Images

Raw microarray data first exists as a scanned image file. Differences in spot size, non-uniformity of spots, heterogeneous backgrounds, dust and scratches all contribute to variations at the image level. In Chapter 5, David Lalush characterizes such parameters and discusses ways to simulate additional microarray images for use in developing image analysis algorithms.

On the Affymetrix platform, hybridization operators have observed that the images tend to form some kind of mysterious pattern. In Chapter 6, Andrey Ptitsyn argues that there indeed is a background pattern. He further postulates that the pattern might be caused by the fluid dynamics in the hybridization chamber.

Normalizing Raw Data

Normalization has been recognized as a crucial step in data pre-processing. Do some mathematical operations truly allow us to remove the systematic variation that might skew our analysis, or, are we distorting the data to create illusions? The paper by Liling Warren and Ben Lu. investigated seven different ways to normalize microarray data. Results show that normalization has a greater impact than expected on detecting differential expressions: the same downstream detection method can result in 23 to 451 genes, depending on the pre-processing of the data. Suggestions to guide researchers in the normalization process are provided.

Characterizing Technical and Biological Variance

The project normal paper [PNAS 98:13266-77, 2001] showed us that even for animals under 'normal' conditions, gene expression levels do fluctuate from one to the other. This biological variation complicates the final genetic variation we find on the microarray. The microarray can also include technical variations produced during the measurement process. Deng *et al.* describes a two linear mixed model to assess variability and significance in Chapter 8. By a similar mixed model approach, Cui *et al.* calculates the necessary number of replicates to detect certain changes. This is of great interest to experimental biologists. Usually, we have limited resources for either total number of microarrays as a financial consideration, or from the limited number of cells we can obtain. The optimal resource allocation formula by Cui *et al.* lets us answer questions such as: Should we use more mice or more arrays? Should we pool mice? Chapter 9 provides some answers to these questions.

Estimating location and scale from experimental measurements has been one of the major themes in statistics. Most of the previous work on microarrays focused on the classification of the expression changes under different conditions. Bhattacharjee *et al.* investigated the classification of intrinsic biological variance of gene expression. By using a Bayesian framework, the authors support the hypothesis that some genes by nature exhibit highly varied expression. This work is featured in Chapter 10.

Investigating Cross Hybridization on Oligonucleotide Microarrays

Quantitative binding of genes on the chip surface is a fundamental issue of microarrays [Nature Biotechnology 17:788-792, 1999]. Characterizing specific versus non-specific binding on the chip surface has been important yet under studied. The Affymetrix Latin Square data set provides an excellent opportunity for such studies. All three papers in the next section of this volume address this issue. In 2002, Zhang, Miles, and Aldape developed a free energy model of binding on microarrays [Nature Biotechnology 2003, In Press]. This mechanistic model is a major development since the Li-Wong statistical model [PNAS 98:31-36, 2001] for hybridization. At CAMDA '02, Zhang *et al.* demonstrated an application of this model to identify spurious cross-hybridization signals. Their work is highlighted in Chapter 11. Kachalo *et al.* suggest in Chapter 12 that a match of 7 to 8 nucleotides could potentially contribute to non-specific binding. This non-specific binding provides clues to the interpretation of hybridization results but also assists in the future design of oligos. The same cross-hybridization issue also caught the attention of statisticians. Hsieh *et al.* exposed cross-hybridization problems by studying outliers in the data set. In this case, the cause of cross-hybridization was found to be fragments matching the spiked-in genes. In summary, these three papers expound on potential cross-hybridization problems on microarrays and suggest some solutions.

Finding Patterns and Seeking Biological Explanations

The final section of this volume focuses on utilizing the Gene Ontology (GO) as an explanatory tool, though the final two papers differ in how to group the genes. Moloshok *et al.* modified their Bayesian decomposition algorithm to identify the patterns in gene expression and to specify which gene belongs to which pattern. The Bayesian framework not only allows encoding prior information in a probabilistic way, but also naturally allows genes to be assigned to multiple classes. In the final chapter, Diaz-Uriarte *et al.* use GO to obtain biological information about genes that are differentially expressed between organs in the Project Normal data set. The techniques incorporate a number of statistical tests for possible identification of altered biochemical pathways in different organs.

Summary

The CAMDA '02 conference again brought together a diverse group of researchers who provide many new perspectives to the study of microarray data analysis. At previous CAMDA conferences we studied expression patterns of yeast and cancers. The most recent CAMDA conference took a step back to find variations in the data and possible problems. Our next

CAMDA conference ('03) will focus on data acquired at different academic centers and the problems in combining that data.

Web Companion

Additional information for many of these chapters can be found at the CAMDA website, where links to algorithms, color versions of several figures, and conference presentation slides can be found. Information about future CAMDA conferences is available at this site as well. Please check the website regularly for the call for papers and announcements about the next conference.

http://camda.duke.edu

SECTION I

TUTORIALS

1

THE BIOLOGY BEHIND GENE EXPRESSION: A BASIC TUTORIAL

Michael F. Ochs* and Erica A. Golemis
Division of Basic Science, Fox Chase Cancer Center, Philadelphia, PA

Abstract: Microarrays measure the relative levels of gene expression within a set of cells isolated through an experimental procedure. Analysis of microarray data requires an understanding of how the mRNA measured with a microarray is generated within a cell, how it is processed to produce a protein that carries out the work of the cell, and how the creation of the protein relates to the changes in the cellular machinery, and thus to the phenotype observed.

Key words: Transcription, translation, signaling pathway, post-translational modification

1. INTRODUCTION

This tutorial will introduce key concepts on cellular response and transcriptional activation to non-biologists with the goal of providing a context for the use and study of microarrays. The field of gene regulation, including both transcriptional control (the regulation of the conversion of DNA to messenger RNA) and translational control (the regulation of the conversion of messenger RNA to protein), is immense. Although not yet complete, our understanding of these areas is deep and this tutorial can only touch upon the basics. A deeper understanding can be gained through following the references given within the text, which are focused primarily on recent reviews, or through one of the standard textbooks, such as Molecular Biology of the Cell [Alberts *et al.*, 2002]. It should be noted that we are focusing on eukaryotic animal organisms within this tutorial, and not all aspects of the processes discussed will apply to plants or prokaryotes, such as bacteria.

* author to whom correspondence should be addressed

The typical microarray experiment aims to distinguish the differences in gene expression between different conditions. These can be between different time points during a biological process, or between different tissue or tumor types. While standard statistical approaches can provide an estimate of the reliability of the observed changes in the messenger RNA levels of genes, interpretation and understanding of the significance of these changes for the biological system under study requires far more detailed knowledge. Of particular importance is the issue of nonlinearity in biological systems. For any analysis beyond the simple comparison of measurements between two conditions, the fact that the system being studied is a dynamic nonlinear system implies that the analysis must take into account the interactions present in the system. This is only possible if the system is understood at a non-superficial level.

The nomenclature surrounding the biological structures and processes involved in transcription and translation is substantial. The following table contains a glossary of terms that can be referenced as needed.

Table 1. A glossary of key terms used for describing cellular processes of transcription and translation. For human genes and proteins, the convention is to use capital italics (e.g., *BRCA1*) for the gene and capitals for the protein (e.g., BRCA1). Unfortunately the conventions are organism specific.

GLOSSARY	
Transcription	
DNA	deoxyribonucleic acid, a polar double-stranded molecule used for the fundamental storage of hereditary information and located in the nucleus (polarity is defined as 5' – 3')
GATC	guanine, adenine, thymidine, cytidine: the 4 "bases" that comprise the DNA strands
gene	fundamental DNA unit for production of a protein, transcribed along a specific DNA sequence
transcription	the process by which a gene encoded in DNA is converted into single-stranded pre-mRNA
RNA	ribonucleic acid, a single stranded molecule with multiple uses within the cell
promoter	a region of DNA proximal to the 5' end of a gene, where proteins bind to initiate transcription of a gene
enhancer	a region of DNA where binding of transcriptional regulatory proteins alters the level of transcription of a gene
transcriptional activator	a protein or protein complex that binds to a promoter or enhancer to induce transcription of a gene
transcriptional suppressor	a protein or protein complex that binds to a promoter or enhancer to block transcription of a gene
chromatin	higher-order compaction structure for DNA, based on assembly of DNA around histones in nucleosomes
chromatin remodeling	a process wherein the phasing of nucleosomes is altered, altering the access of transcription factors to the DNA

GLOSSARY	
histones	proteins that provide structural elements for the creation of chromatin
coactivator	proteins that modify histones allowing chromatin remodeling
basal transcriptional apparatus	a protein complex that binds to a transcriptional initiation site and performs conversion of DNA into mRNA
mRNA	messenger RNA, a form of RNA created from DNA generally containing a 5' untranslated sequence, a protein coding region, and a 3' untranslated sequence
rRNA	ribosomal RNA, a form of RNA that associates in a complex with ribosomal proteins, which constitute the structural elements of the ribosome
tRNA	transfer RNA, the form of RNA that recognizes individual codons on an mRNA and brings the amino acid specified by that codon for incorporation into a polypeptide
codon	a three nucleotide sequence (RNA or DNA) that specifies a specific amino acid to insert in a protein
mRNA – associated terms	
exon	a portion of genomic DNA encoding a sequence of amino acids to be incorporated in a protein (coding region)
intron	a portion of genomic DNA that does not code for a sequence of amino acids to be placed in a protein, but intervenes between exons within a gene (noncoding region)
pre-mRNA	the mRNA polymer that is initially transcribed from DNA, which contains both exons and introns
splicing	the process by which an mRNA is modified to remove introns and sometimes vary the exons included in the mRNA prior to translation
poly-adenylation	the addition of approximately 20-50 adenine residues to the end of an mRNA sequence
splice variants	different mRNAs created from the same DNA sequence through different splicing
mRNA export receptor	a protein that transports the mRNA from the nucleus to the cytoplasm
mRNA export sequence	a portion of mRNA that is recognized by an export receptor allowing the mRNA to be exported from the nucleus
miRNA	micro-RNA, a short RNA (<22 nucleotides) encoded within the genome that can bind to a specific mRNA and block its translation into a protein or alter its stability
siRNA	small interfering RNA, a short, artificial RNA designed to bind a specific mRNA and block its translation into a protein, this is now widely used for gene knockout studies
Translation	
amino acid	one of 20 specific molecules used within all living creatures to construct proteins, containing a common core (N-C-COOH) and varying sidechains attached to the central carbon

GLOSSARY	
amino acid residue	the term used to refer to an amino acid following insertion within a polypeptide chain
polypeptide chain	a series of linked amino acid residues produced during the process of translation
protein	a completed polypeptide chain that folds into a three dimensional structure to provide a cellular function specified by a gene
ribosome	a complex of proteins and rRNA that builds a polypeptide chain based on the codons in an mRNA
chaperone	a large molecular machine that assists a polypeptide chain in proper folding into a protein
Proteins	
dimer	a small complex formed by the non-covalent association of two proteins
homodimer	a dimer that contains two identical subunits (i.e., is made from two copies of the same protein)
heterodimer	a dimer that contains two different subunits (i.e., is made from two different proteins)
kinase	a protein that adds a phosphate group to specific amino acid residues in a protein in a process called phosphorylation
phosphatase	a protein that removes a phosphate group from specific amino acid residues in a protein
scaffolding protein	a protein that binds multiple other proteins into a specific conformation, enhancing or otherwise controlling their interactions
ubiquitination	the addition of one or more copies of ubiquitin (a short peptidyl sequence) to a protein, either targeting a protein for degradation or contributing to control of protein localization
proteasome	a very large protein complex that degrades ubiquitinated proteins
Signaling	
signal	an abstraction of protein modifications indicating the transmission of information within a cell
signal transduction	the use of programmed, frequently sequential changes in protein interaction, modification, and activity status to transmit information within a cell in a "signaling cascade"
signaling activator	a protein that modifies a second protein activating the function of that protein
signaling inhibitor	a protein that modifies a second protein suppressing the function of that protein
receptor	a membrane bound protein that can receive a signal of extracellular origin to activate a signaling cascade
ligand	a small protein, peptide, or hormone that binds to a cognate receptor to initiate signaling
signaling pathway	a group of proteins that provide the physical mechanism for a signaling cascade, associated with a specific biological response

GLOSSARY	
junction	a point where multiple signals combine and can be integrated
node	a point where a single signal diverges and can provide input into multiple downstream points
signaling network	a group of signaling pathways linked together at junctions and nodes creating a nonlinear response system
Cellular Structures	
membrane	a lipid bilayer separating compartments, such as the outer cell membrane that separates a cell from its environment
nucleus	a region separated from the rest of the cell by a membrane and containing the DNA
cytoplasm	the portion of the cell outside the nucleus and containing numerous smaller structures
ER, endoplasmic reticulum	a membrane-based cellular structure that serves multiple functions including localization of some ribosomes, and processing of membrane-associated and secreted proteins
Golgi apparatus	a cellular structure that receives proteins following passage through the ER, providing for additional protein modification and transport
cytoskeleton	the complex of multiple structural proteins (e.g., actin, tubulin) that provide structural integrity to the cell
ECM, extracellular matrix	the collection of structural proteins secreted by cells, and forming an external "mesh", that contributes to cell shape control, survival functions, and signaling processes

2. CELLULAR RESPONSES AND SIGNALING PATHWAYS

In order to survive, grow, and interact in complex organisms, cells must interpret signals coming from both the external and internal environments. Most behaviors, and all complex responses, require signaling pathways that encode within the cell the control mechanism for response behaviors.

A simple example of such a system is shown in Figure 1. The yeast *Saccharomyces cerevisiae* has developed a signaling pathway that responds to external signals in order to initiate a mating response [Posas *et al.,* 1998]. The external signal is detected by a transmembrane protein, the receptor Ste2p. When a ligand specific for Ste2p produced by a cell of a different mating type binds to this receptor, a signaling cascade is induced within the cytoplasm of the yeast cell. The conformational change induced by the ligand binding first leads to the cleavage and activation of a G protein, which in turn activates Ste20p, a protein kinase. Ste20p in turn activates Ste11p, which activates Ste7p, which activates Fus3p, which activates a transcription factor, Ste12p. The transcription factor then initiates transcription within the nucleus leading to the creation of messenger RNA (mRNA). Only the end result of this cascade is measured with a microarray.

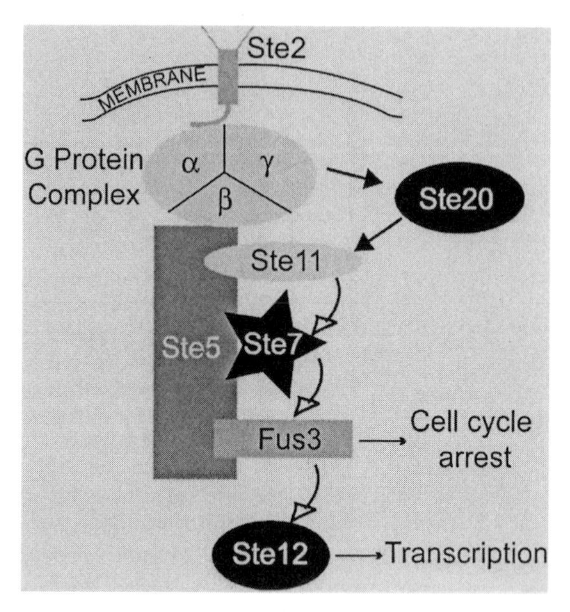

Figure 1. A simple signaling pathway. The signaling pathway for the mating response in budding yeast is shown schematically. The response is triggered by binding of a ligand to the Ste2 membrane receptor, which induces sequential activation of the cascade down to Ste12, a transcription factor. Ste5 is a scaffolding protein that binds the Ste11, Ste7, and Fus3 proteins in a complex, increasing the efficiency of the signaling cascade.

A set of proteins functionally linked in a single pathway is described as a signaling cascade or pathway, with the process of movement of information through a signaling cascade described as signal transduction. It is immediately obvious that the cell is going to a great deal of trouble to respond to a simple signal, introducing several intermediaries between signal (ligand binding) and response (transcription). However, one reason for this complexity is that it allows encoding of complex behavior by the creation of signaling networks [Jordan *et al.,* 2000]. A signaling network is created by the intersection of multiple signaling pathways, with these intersections generally occurring at multiple points within each separate pathway. The result is a highly nonlinear system capable of responding in multiple ways to a signal depending on the state of the overall network and the presence of signals affecting different pathways.

The structure of a representative network is shown in Figure 2. This is a highly simplified portrayal of an important pathway in human cancer involving the cancer-inducing gene *RAS* (italics for a gene); see the review by Kolch for a full discussion [Kolch, 2000]. In normal cells, RAS (no italics for a protein) is activated by an external stimulus through the epidermal growth factor receptor and other growth factors, and then it interacts with numerous other proteins to determine a cellular response. There are multiple, distinct receptors that may be activated in response to different stimuli, and each receptor may activate multiple pathways. The network is comprised of junctions, where multiple signals come together (e.g., RAF and RAC in Figure 2), and nodes, where a single signal can branch to multiple pathways (e.g., AKT and RAS in Figure 2). In addition, for any component within a pathway, there can be both activators (often kinases) and inhibitors (often phosphatases). For this diagram, the inhibitors and activators are not separated. Additionally, there is generally feedback within the network, either through proteins created as a result of the transcriptional activators (circles in Figure 2) or directly from loops in the network (AKT to RAF together with AKT to RAC to PAK to RAF). A junction can therefore represent a point where the result of one pathway's activation is the activation of a kinase, while the result of a second pathway's activation is activation of a phosphatase, both targeted at the same protein. A node may be a kinase that phosphorylates multiple proteins.

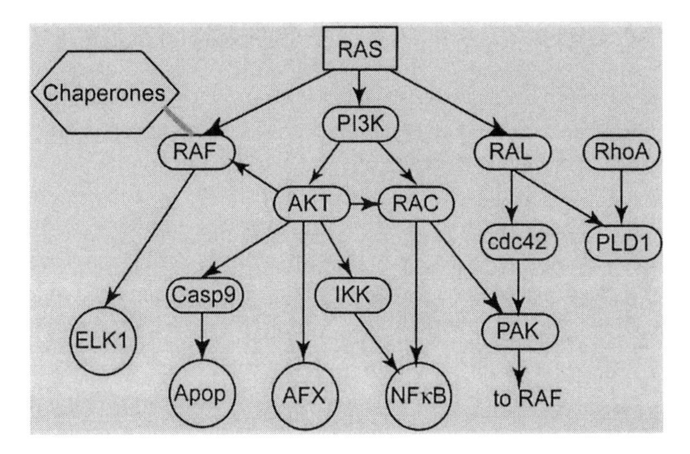

Figure 2. A simplified signaling network. The RAS/RAF pathway is a critical pathway in the development of many cancers. The interactions are complicated and can be either inhibitory or activating. In addition, feedback loops are common, both within the intermediate signaling proteins (e.g., there are multiple paths to RAF) and through the transcription factors (circles), which produce mRNAs encoding proteins that can feed back signals to the pathways. The factor indicated as Apop is actually a cascade leading to apoptotic signals (which can cause cell death) and transcription of pro-apoptotic genes. A number of intermediate steps have been left out of the pathways shown.

In order for kinases and phosphatases to operate, they must be in close proximity to the target protein. This adds an additional degree of complexity, as cellular localization becomes a key issue. While some localization signals are encoded by the signaling proteins themselves, a set of proteins, known as scaffolding proteins (e.g., Ste5 in Figure 1), has evolved to keep the necessary components of a signaling pathway together [Tzivion *et al.,* 2001]. Scaffolding proteins may not always be absolutely necessary for signal transduction, however they increase the efficiency of signaling. And, as noted above, a signaling network is a complex nonlinear system, so that signal strength can play a major role in the outcome of signal initiation.

As multiple signals arrive at junctions, the response of that junction will depend on the relative strength of various signals and the present overall state of the cell (reflected as levels of kinase and phosphatase activity in the local environment, protein interaction profiles, etc.). Each junction then transduces a signal, until a physiological response occurs. This response can include changes in cellular movement, intracellular transport or metabolism, protein state (through degradation or modification), or transcriptional and translational responses. While transcriptional activation is of interest here, it is important to note not only that it is only one of a myriad of potential changes occurring after activation of signaling, but also that transcription is generally a late, downstream indicator of activity.

3. TRANSCRIPTION

Since gene expression is a downstream indicator of activity, it is important to understand the basics of how the mammalian cell converts its 3,000,000,000 bases of DNA into usable, small messenger RNA (mRNA). Within cells, DNA is organized into chromatin. The double-stranded DNA is wound into coils around protein complexes, composed of proteins called histones. While it has long been understood that chromatin must be remodeled (i.e. unwound and rewound in a different pattern) as part of initiating transcription, it has only lately become clear that the histones play a larger role in regulation of expression [Berger, 2002]. There are numerous proteins that modify histone structure, through mechanisms including acetylation, methylation, phosphorylation, and ubiquitination of the histone proteins. The proteins that modify histones, called coactivators, are generally recruited to the DNA by transcriptional activators, which bind the DNA directly [Featherstone, 2002]. Transcriptional activators bind the

DNA at specific sequences, called promoters and enhancers. Promoters lie upstream of the gene, usually within a few kilobases of the start site, while enhancers can occur both upstream and downstream from the gene and can lie farther from the gene itself.

The overall machinery governing transcription is still an area of very active study. The signal to transcribe a gene is first transduced to the nucleus through a signaling pathway activating a transcription factor. This can be done directly by proteins such as nuclear receptors, which are capable of directly entering the nucleus and initiating transcription [Dilworth *et al.*, 2001] following their activation in the cytoplasm, or indirectly by propagation of signals through a chain of protein interactions culminating in activation of a nuclear transcription factor. The transcription factor must then activate the basal transcriptional apparatus, which includes the key TATA-box binding protein and RNA polymerase II [Green, 2000]. Once the transcriptional activator is active in the nucleus, it binds to DNA at its promoter. The transcription factor then recruits the coactivators, which leads to chromatin remodeling, and activates the transcriptional apparatus, following which transcription begins. The transcription process is itself complex, requiring unwinding and opening of the DNA and translocation of the complex along the DNA [Reines *et al.*, 1999], but we will not detail these issues here.

While much is shared between prokaryotic organisms, such as bacteria, and eukaryotic organisms, such as ourselves, there are significant differences in the transcriptional apparatus and structure of DNA. Within prokaryotes the genes are arranged on the chromosomes in a linear fashion, each gene comprised of a continuous DNA sequence that is transcribed into a corresponding mRNA sequence. There tends to be only short sequences of DNA between coding regions, and these sequences include the promoters. Within higher eukaryotes, a gene generally contains both introns (sequences which are not contained in the final mRNA) and exons (sequences that remain in the final mRNA). Furthermore, the intergenic DNA becomes significantly larger, with tens to hundreds of kilobases between transcribed units, and individual genes spread over considerable distances. The mRNA of a gene is transcribed first as a long sequence containing both introns and exons, and then spliced to a continuous message comprised only of exons [Sharp, 1994]. This introduces the possibility of variant proteins encoded within a single "gene" in the DNA through splice variants, where different exons are combined to produce a final mRNA transcript.

The spliced mRNA must then be transported out of the nucleus for translation into a polypeptide. This is done through a highly conserved export receptor that appears to be coupled to the mRNA splicing machinery upstream [Reed *et al.*, 2002]. The export mechanism is not well understood, but it appears that mRNAs have multiple export paths that depend on specific export sequences encoded within the mRNA [Stutz *et al.*, 1998].

An additional step in the transcriptional machinery was discovered recently. The existence of small lengths of RNA capable of silencing the translation of mRNA were first noted in plants, and later confirmed to be present in all organisms [McManus *et al.*, 2002]. Essentially, functional codes for extremely short lengths of RNA appear to be part of our genetic structure. These micro RNAs (miRNAs) get converted within our cells to approximately 21 base-pair single stranded RNA units that complement sequences in the noncoding regions of mRNA. The binding of the single stranded RNA to the mRNA targets the mRNA for destruction. Effectively this "silences" the gene by blocking translation.

4. TRANSLATION

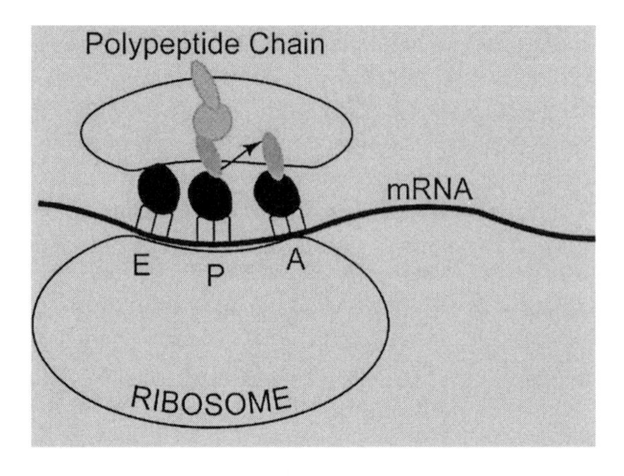

Figure 3. The translational apparatus. The processed mRNA is transported to the ribosome, where it is translated into a polypeptide chain. The tRNAs bind to the ribosome, the anticodon is checked against the mRNA codon, and the peptide chain grows as each tRNA contributes its amino acid to the chain. The tRNAs are represented by the black ovals with three feet representing the bases of the anti-codon, while the amino acids are represented by the gray ovals.

Following export, mRNA must then be transported to the ribosomes for translation into a protein. Ribosomes are large complexes made from ribosomal RNA (rRNA), another form of RNA present within the cell, and ribosomal proteins. Ribosomes comprise two subunits that clamp onto the mRNA chain and process it in a linear fashion [Ramakrishnan, 2002], as shown in Figure 3. Within an mRNA, "triplets" of nucleotides, termed codons, are arranged in sequence to specify the amino acids and their order for the protein encoded by the gene. The ribosome contains three motifs that bind transfer RNA (tRNA), yet another form of RNA existing within the cell. Each tRNA comprises an "anticodon", three bases keyed to bind three mRNA bases, and a structure to bind a specific amino acid. The three binding sites on the ribosome, labeled A, P, and E, position the tRNAs to test a match to the mRNA template, then transfer a growing polypeptide chain, and release the tRNA. The tRNA is first matched to the mRNA at the A site. The existing polypeptide chain that is bound to the tRNA at the P site is transferred to the amino acid on the tRNA on the A site, then this tRNA translocates to the P site, while the tRNA on the P site translocates to the E site. It is released as the next tRNA binds to the now vacant A site. In this way, the ribosome builds a protein translated from the codons encoded in the mRNA.

When completed, the ribosome releases a polypeptide chain that must then fold to become a functional protein. Protein folding generally involves a chaperone, a molecular machine which aids the polypeptide chain in folding into the correct conformation [Zhang *et al.*, 2002]. If the folding fails to produce the correct structure, the cell has ways to target the protein for degradation as well as to unfold and refold the protein. Finally, the protein often requires transport to the correct cellular compartment. For instance, membrane receptors must be transferred from the ribosomes to the membrane; nuclear proteins must be moved from the cytoplasm into the nucleus.

The translation of the proteins can occur in multiple locations. Generally, proteins that will remain in the cell are translated by ribosomes in the cytosol, while proteins to be secreted are produced in the endoplasmic reticulum, processed through the Golgi apparatus, and rapidly exported from the cell.

5. PROTEIN ACTIVITY

Once the mRNA has been translated into a protein, the subsequent processing, life expectancy, profile of modifications, and means of function of the proteins can be extremely complex, and differs greatly from protein to protein. As proteins accomplish the effective "work" of a cell, understanding the points necessary to insure creation of a competent work "unit" is necessary to begin to think about how expression changes seen in a microarray can result in functional consequences for a cell or organism. For some proteins, primarily those involved in metabolism, this can be straightforward. For most proteins, it is complex.

Protein stability varies considerably following translation. Some proteins have extremely long half-lives, of the order of 20 hours or more. Other proteins have extremely short half-lives (~2-3 minutes). This difference in stability reflects the different biological roles of these proteins, which in some cases require activity as a stable structure (for instance, as part of the cytoskeleton providing cellular architecture), but in other cases require rapid turnover (as with proteins critical for execution of a chronologically limited step in cell cycle progression). Further, some proteins exist at different locations within cells, or in association with different partner proteins: based on their location or pattern of association, some intracellular pools of a given protein may be subject to more rapid degradation than others. These differences in control of the lifespan of proteins add to the difficulty in extrapolating protein levels from mRNA levels.

In addition to changes in protein levels, the majority of signaling proteins are subject to extensive post-translational modifications that can affect their stability, localization, and activity. For example, the NF-κB protein initially exists as an inactive precursor cytoplasmic 105 kD (kiloDalton, a measure of mass) form that is modified by covalent attachment of ubiquitin, and then processed to release a 50 kD form that transits to the cell nucleus and activates transcription. A number of other important proteins are similarly cleaved and relocated. Another common form of modification is the attachment of small peptidyl or lipid moieties (e.g., SUMO; myristylation) that again control patterns of localization, association, and stability. However, by far the most common form of protein modification is phosphorylation. The placement of one or more phosphate groups on proteins by one or more kinases, and the subsequent removal of these groups by phosphatases, can have drastic effects on every aspect of protein function, and lies at the core of studies of signal transduction. A rapid screen through the CGAP signaling resource sponsored by the NIH* emphasizes the prevalence of this form of control. Again, different pools of a protein will be differentially phosphorylated, and hence differentially active against different targets, within a cell. For some well-studied proteins it is barely possible to begin to estimate what percentage of an expressed protein pool is active following a given set of stimuli; for the majority, it is currently an open question.

Another complicating factor in understanding the activity level of a protein is the fact that many proteins function for part or all of their existence as components of complexes involving other proteins. Some transcription factors only gain their specificity in binding DNA if they heterodimerize with other proteins. The majority of proteins involved in

*http://cgap.nci.nih.gov/Pathways

signal transduction become activated through association (sometimes sequential, sometimes simultaneous) with a large number of other proteins. Further individual proteins may associate with different sets of partners to have separate activities in different signaling cascades. Hence, the expression of one protein, A, in a cell may well have no functional consequences for output "X", if the requisite partners for activity in the X pathway are absent, but function well for output "Y", because partners required for that activity are present. In some cases, these requisite partners may be co-transcriptionally induced with A, and hence predictable by microarray. In other cases, the partners may pre-exist either ubiquitously (in all cell types) or selectively (in some cell types), and not be detectable by methodologies tracing transcriptional induction. Based on two-hybrid or mass spectrometry proteomics efforts using yeast as a model system, it is clear that the majority of proteins are engaged in many interactions, with current estimates suggesting in excess of 10, for any given protein. This is likely to be an underestimate.

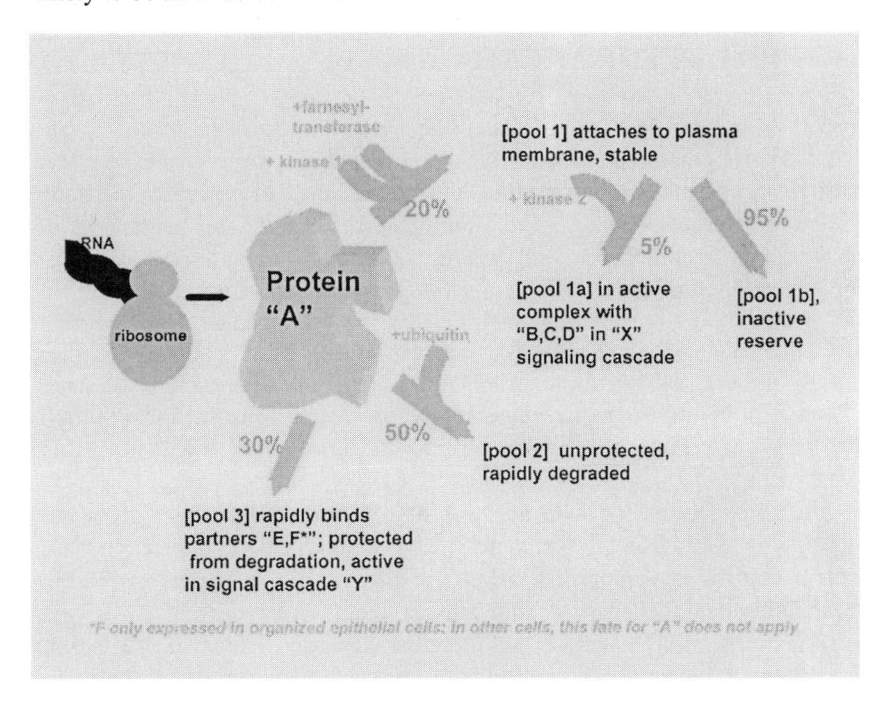

Figure 4. Hypothetical post-translational processing of a protein. There are many modifications that proteins undergo after translation which affect their activity. A number of possible fates are shown for a single protein, and these fates are often dependent on cell type.

These points of control are briefly summarized in Figure 4. In fact, this description represents a considerable over-simplification of factors regulating protein activity. As one example of interest, it has long been known to cell biologists that cells can have very different responses to specific growth stimuli when on supports that allow them to form three-dimensional organized structures rather than on a flat tissue culture plate. In a number of cases, an appropriate 3-D structure has been shown to be required for expression of an appropriate transcriptional program [DiPersio *et al.,* 1991]. In other cases, formation of appropriate cell-cell contacts can control cellular resistance to drugs and sensitivity to apoptosis [Jacks *et al.,* 2002], in a process that appears to involve regulation of the availability and functionality of signaling proteins and transcription factors. As a key issue in all microarray work is the degree to which *in vitro* (i.e., cell culture) and *in vivo* (e.g., tumor) data can be compared, it is important to keep such higher order regulatory mechanisms in mind.

6. ISSUES FOR MICROARRAYS

As the above sections indicate, the cell is a complex machine that evolution has guided to allow it to respond to external threats, survive under multiple conditions, and, in multicellular organisms, cooperate for the good of the greater organism. Evolution occurs through the borrowing of function, the recombination of existing functions into new ones, and the copying and modification of existing functions. The result is that proteins, the primary functional components, have multiple roles and multiple partners allowing the cell to vary the response to a stimulus based on other stimuli, the external environment including other cells, and on internal state. Effectively, the cell is a state machine whose response to identical stimuli can vary according to variables beyond the control of any conceivable experiment.

The implications for data analysis of microarray data are significant. First, as is clear from the complexities of transcription, translation, and protein activity, it is a hopeless task to use changes in the level of expression of an individual gene as an indicator of changes in the activity of the corresponding protein without additional information [Chen *et al.*, 2002]. Essentially, the changes in expression are not upstream indicators of protein activity but instead comprise downstream indicators of changes in the signaling pathway and cell state.

Second, the signaling networks present within cells allow the cell to respond differently to the activation of a given pathway, so that the response to a signal will vary depending on cellular state and the external environment. Cellular state is impacted by factors such as circadian rhythm and metabolic oscillations that cannot be well controlled in the laboratory.

Third, the destruction of mRNA can occur in multiple ways. Each of the mRNA species has a typical half-life, which varies between species, and can also be targeted for degradation by small RNAs. These issues lead to the relative concentrations of mRNA being dependent on the timing of mRNA harvest for some subset of the species within a cell.

Fourth, it is clear that the cell encodes multiple "RNA genes" within a single "DNA gene" through mRNA splicing. Depending on the specific sequence spotted onto an array or grown onto a GeneChip, the hybridization recorded may represent only one of these variants or some or all of these variants. However, splice variation changes function just as protein modification does.

Together these biological realities greatly complicate the analysis of microarray data. Essentially, downstream gene expression is highly correlated, with pathway activation leading to multiple genes being expressed, so that expression levels of different genes are not independent. In addition, each set of genes will involve genes included in multiple functions, and that therefore respond to multiple stimuli, so that clusters of genes cannot easily be linked to pathway or function.

Since the cellular responses also vary with cellular state, there is a variability linked to unobserved variables. The variation will unfortunately therefore not be stochastic, but instead will represent an underlying systematic variation within the data. With an adequate number of data points, the nonstochastic nature should become obvious, however present microarray studies are data poor in this regard.

For many microarray studies, the situation is further complicated by the heterogeneic nature of many *in vivo* samples. In general, it is not feasible to obtain pure tissue types from tumors or other tissues. The measurements made on a microarray then represent expression arising from multiple tissue types (e.g., tumor and surrounding "normal" tissues) rather than from only the tissue of interest.

In summary, biological systems exhibit complex, highly regulated behavior with significant feedback in all aspects and at all points in the response to stimuli, from detection of a signal through activation of a response to creation of the means to respond to the stimuli. Such systems include the ability to block the response at multiple stages (e.g., signal transduction, transcription, translation, protein activation) and to signal between each stage. This makes the systems highly nonlinear and prone to the creation of highly correlated errors in derived data. Analysis of the data should not go forward blind to these realities.

ACKNOWLEDGMENTS

We thank the National Institutes of Health, National Cancer Institute (CCCG CA06927 to R. Young), the Pennsylvania Department of Health (grants to mfo and eag), and the Pew Foundation for support.

REFERENCES

Alberts, B, Lewis, J, Raff, M, Johnson, A, and Roberts, K (2002) Molecular Biology of the Cell, 4th Ed., Taylor and Francis, Inc., London.

Berger, SL (2002) Histone modifications in transcriptional regulation. Curr Opin Genet Dev 12: 142-8.

Chen, G, Gharib, TG, Huang, CC, Taylor, JM, Misek, DE, Kardia, SL, Giordano, TJ, Iannettoni, MD, Orringer, MB, Hanash, SM and Beer, DG (2002) Discordant protein and mRNA expression in lung adenocarcinomas. Mol Cell Proteomics 1: 304-13.

Dilworth, FJ and Chambon, P (2001) Nuclear receptors coordinate the activities of chromatin remodeling complexes and coactivators to facilitate initiation of transcription. Oncogene 20: 3047-54.

DiPersio, CM, Jackson, DA and Zaret, KS (1991) The extracellular matrix coordinately modulates liver transcription factors and hepatocyte morphology. Mol Cell Biol 11: 4405-14.

Featherstone, M (2002) Coactivators in transcription initiation: here are your orders. Curr Opin Genet Dev 12: 149-55.

Green, MR (2000) TBP-associated factors (TAFIIs): multiple, selective transcriptional mediators in common complexes. Trends Biochem Sci 25: 59-63.

Jacks, T and Weinberg, RA (2002) Taking the study of cancer cell survival to a new dimension. Cell 111: 923-5.

Jordan, JD, Landau, EM and Iyengar, R (2000) Signaling networks: the origins of cellular multitasking. Cell 103: 193-200.

Kolch, W (2000) Meaningful relationships: the regulation of the Ras/Raf/MEK/ERK pathway by protein interactions. Biochem J 351 Pt 2: 289-305.

McManus, MT and Sharp, PA (2002) Gene silencing in mammals by small interfering RNAs. Nat Rev Genet 3: 737-47.

Posas, F, Takekawa, M and Saito, H (1998) Signal transduction by MAP kinase cascades in budding yeast. Curr Opin Microbiol 1: 175-82.

Ramakrishnan, V (2002) Ribosome structure and the mechanism of translation. Cell 108: 557-72.

Reed, R and Hurt, E (2002) A conserved mRNA export machinery coupled to pre-mRNA splicing. Cell 108: 523-31.

Reines, D, Conaway, RC and Conaway, JW (1999) Mechanism and regulation of transcriptional elongation by RNA polymerase II. Curr Opin Cell Biol 11: 342-6.

Sharp, PA (1994) Split genes and RNA splicing. Cell 77: 805-15.

Stutz, F and Rosbash, M (1998) Nuclear RNA export. Genes Dev 12: 3303-19.

Tzivion, G, Shen, YH and Zhu, J (2001) 14-3-3 proteins; bringing new definitions to scaffolding. Oncogene 20: 6331-8.

Zhang, X, Beuron, F and Freemont, PS (2002) Machinery of protein folding and unfolding. Curr Opin Struct Biol 12: 231-8.

2

MONITORING THE QUALITY OF MICROARRAY EXPERIMENTS

Kevin R. Coombes, Jing Wang, Lynne V. Abruzzo
University of Texas M.D. Anderson Cancer Center, Houston, TX.

Abstract: A microarray experiment is a complex, multistep process involving biology, chemistry, physics, and bioinformatics. Something can go wrong at every step in the process. In order to obtain good results, one needs a thorough, redundant system to monitor the quality of microarray experiments. In this article, we provide an overview of quality control measures that can be applied at different points during the process of conducting and analyzing microarray experiments.

Key words: microarrays, quality control, process control, acceptance testing

1. INTRODUCTION

A microarray experiment is a complex, multistep process. Clones of known DNA sequences must be grown, harvested, and spotted in precise locations on the microarray substrate. RNA must be extracted from experimental samples, labeled targets must be produced by reverse transcription, and the targets must be hybridized with a microarray. The hybridized microarray must be scanned to produce a computer image. The image must be loaded into a software package for quantification, a template grid must be precisely aligned, and estimates must be computed for spot intensity and local background. The initial intensity estimates must be background-corrected and normalized. In a project that includes tens or hundreds of individual microarray experiments, the final intensity estimates from each individual microarray experiment must be combined into a single data structure for further analysis.

Something can go wrong at every step in the process. Every time a person handles a physical microarray, views an image of a microarray in an image-editing program, explores the quantifications in a spreadsheet or database, or saves and labels a file, there is a potential for errors to occur. To obtain the best possible results, one must devise a thorough, redundant system to monitor the quality of microarray experiments.

An illustration of the potential difficulties is provided elsewhere in this volume [Stivers *et al.*, 2003]. In their analysis of the Project Normal data set, Stivers and colleagues found that the UniGene annotations of the spots on the microarrays were inconsistent in the data files supplied for the 2002 conference on the Critical Assessment of Microarray Data Analysis. In response to this finding, the authors of the original Project Normal study [Pritchard *et al.*, 2002] conducted an extremely careful and detailed review of their data quality. During the conference, they confirmed that errors had occurred when merging the quantified expression data into a spreadsheet with the UniGene annotations. They also uncovered minor problems with a small number of quantifications, caused by subgrids that had not been correctly aligned. They prepared a revised data set, which is available from the conference web site (http://www.camda.duke.edu/camda02/contest.asp), thus creating perhaps the most accurate microarray data set currently available.

In this article, we provide an overview of quality control measures that can be applied at different points during the process of conducting and analyzing microarray experiments.

2. QUALITY CONTROL BY BIOLOGISTS

Microarray experiments undergo a critical phase change during the scanning process: they pass from the physical world of clones, plates, slides, and robots into the virtual, computerized world of bioinformatics. During the physical phase, quality control is driven by the biological and chemical properties of the reagents. The most effective method for maintaining a high quality microarray facility is, in one sense, quite simple: decide on a standard set of protocols, validate them, and follow them religiously. Collections of such protocols can be obtained from the Brown lab at Stanford (http://cmgm.stanford.edu/pbrown/protocols/index.html) or from The Institute for Genomic Research (http://www.tigr.org/tdb/microarray/protocolsTIGR.shtml) [Hegde *et al.*, 2001]. During the virtual phase of a microarray experiment, both the source of errors and our ability to detect and control them is governed by bioinformatics. Bioinformatics quality can best be maintained by avoiding spreadsheets or general file naming conventions in favor of a specialized database [Brazma *et al.*, 2002]. Data should be stored and transferred in a format that maintains enough detailed structure to recover biological information about both the samples and the genes on the array [Brazma *et al.*, 2001].

In this section, we describe how to monitor the quality of two critical ingredients—the RNA samples and the microarray slides—before the hybridization step. Because assessment of hybridization quality is typically based on the scanned image, we will reserve its discussion to a later section.

2.1 Monitoring RNA quality

The basic component of a microarray experiment is the RNA extracted from samples. Before running a microarray experiment, this quality should be tested. If enough total RNA is available, it can be evaluated by agarose gel electrophoresis with ethidium bromide. One expects to see crisp bands from the 28S and 18S ribosomal RNA, in a ratio of two-to-one. When the amount of RNA is limited, as is often the case with samples obtained by fine needle aspiration or laser capture microdissection, as little as 5–500 ng total RNA can be evaluated using a model 2100 Bioanalyzer (Caliper Technologies, Mountain View, CA) [Dunmire *et al.*, 2002].

2.2 Monitoring physical array quality

The second fundamental component of a microarray experiment is the physical array itself. In order to obtain good results, the spots on the array should contain uniform, equivalent amounts of spotted polymerase chain reaction (PCR) products or oligonucleotides. Pre-scanning the slide before using it in a hybridization experiment can verify this property. Good results have been reported using nondestructive staining with a fluorescent dye such as SYBR green II [Battaglia *et al.*, 2000] or with SYTO 61 [Yue *et al.*, 2001]. Alternatively, Shearstone and colleagues have described a method for printing oligonucleotide microarrays spiked with small amounts of dCTP-Cy3 or dCTP-Cy5. The slides can be scanned to verify spot morphology by detecting the spiked products, which are then washed off before hybridizing the microarray with the sample of interest [Shearstone *et al.*, 2002].

2.3 Quality control during array manufacturing

Many institutions have established core facilities that manufacture microarrays by spotting either PCR products from cDNA clones or chemically synthesized oligonucleotides on glass slides. Managing, maintaining, and handling large libraries of clones bring their own set of quality control problems. These problems, once again, include a mixture of physical difficulties (contamination from nearby wells; suboptimal conditions for the PCR reactions) and bioinformatical challenges (maintaining the correct annotations of the clones). For example, when we resequenced the cDNA clones being printed on microarrays at the Genomics Core Facility at the University of Texas M.D. Anderson Cancer Center, we found an error rate of 21% in the clones supplied by the manufacturer [Taylor *et al.*, 2001]. Our experience strongly suggests the need to sequence-verify the clones at the final stage before printing them on microarray slides. To date, we have found a much lower error rate among the synthesized oligonucleotides printed on arrays at our facility. Nevertheless, we advise randomly selecting a few oligonucleotides for sequence verification.

Because slides are printed in batches, we can test the quality of an entire batch using established procedures. For example, we can randomly select a small number of slides from each batch. We then test the quality of these microarrays (destructively) by performing a hybridization experiment. If the selected microarrays pass the test, then the batch is accepted.

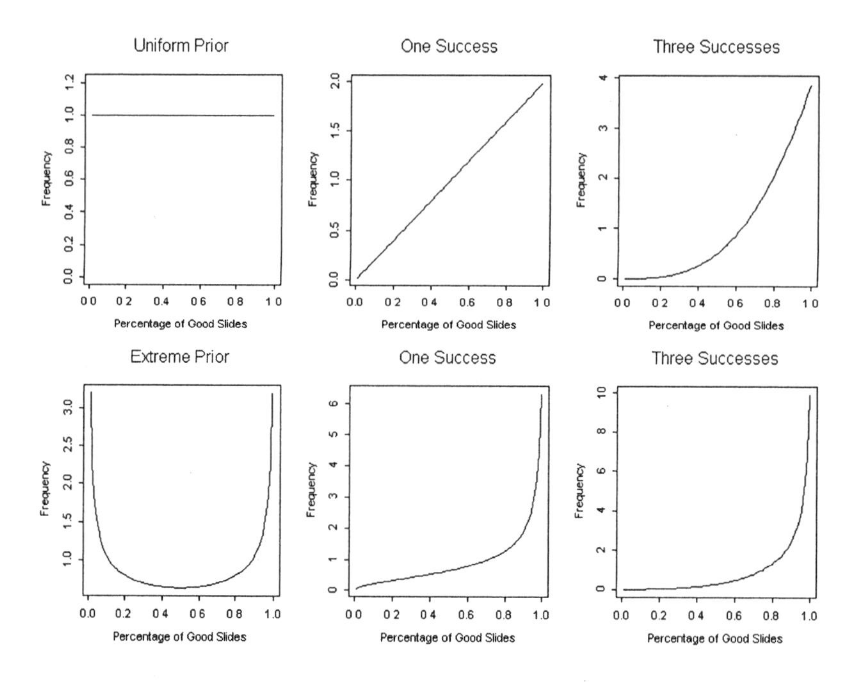

Figure 1. Effects of prior distribution (left panels) on posterior distributions (center, after one array passes the test; right, after three arrays pass the test) of the percentage of good slides in a batch. (Top) Uniform prior. (Bottom) A prior that the batch is likely to be either almost completely bad or almost completely good.

The number of slides that must be tested (and pass) depends on the batch size, the desired confidence level, and other factors. We use Bayesian methods to make this determination. We want to estimate the proportion π of "good" slides in a batch. If we randomly test N slides, then the probability that k of those slides pass the test is described by a binomial distribution

$$P(k \mid \pi, N) = \pi^k (1-\pi)^{N-k} N!/(k!(N-k)!).$$

We assume a Beta prior distribution, $P(\pi) = \text{Beta}(a, b)$. Then standard computations show that the posterior distribution is also a Beta distribution, $P(\pi \mid k, N) = \text{Beta}(a + k, b + N - k)$. The Bayesian formulation allows us to make our assumptions about the prior distribution explicit (Figure 1). For instance, if we believe that a single batch is likely to consist either almost entirely of good slides or almost entirely of bad slides, we can choose prior parameters $a = b = ½$. In this case, testing a single slide can supply enough evidence to determine the quality of the entire batch. If we instead assume a uniform prior on the percentage of good slides, then we will need at least three slides to pass the test in order to accept the batch. One can, of course, make more extensive tests on a few batches to get a better idea of how the quality varies within a batch. This information can be used to make an informed choice of the prior distribution.

There are various choices for the standard hybridization experiment used to test the slides. For instance, all clones spotted on microarrays in our Genomics Core Facility share a common sequence (complementary to the primer). We can label copies of the primer with fluorescent dye and hybridize them to the microarray. We can then assess the attachment of the probes at all spots by scanning the microarrays after hybridization [Hu *et al.*, 2002]. Alternatively, one can chose a standard reference material (such as the Universal Reference available from Stratagene Inc., La Jolla, CA) for this purpose [Weil *et al.*, 2002]. A known profile of this reference material can also be used as part of a process control strategy to monitor the quality of hybridizations over time; we return to this idea later.

3. SPOT LEVEL METRICS

We now turn to quality control methods that apply to individual microarray experiments just after hybridization and scanning. Some of these methods are used to assess the quality of individual spots. Others are used to decide if the overall quality of the hybridization is low enough to reject it and request that the experiment be repeated. In this section, we discuss methods to assess spot quality.

Saturation of the fluorescent signal is one of the more common problems affecting the reliability of spot intensity measurements. Including saturated spots in the analysis can severely distort the results, whether detecting differentially expressed genes or using clustering for class discovery [Hsiao *et al.*, 2002]. Most microarray quantification packages provide a measure of signal saturation. The recommended procedure is to remove frequently saturated gene measurements from consideration. If the number of saturated spots varies widely across experiments, it is usually necessary to revise the scanning protocol in an effort to achieve greater consistency.

Software quantification packages provide different measures of spot quality. Yidong Chen has developed a package to quantify microarray images that includes several spot quality metrics [Chen *et al.*, 2002]. This package computes quality measures for the fluorescent intensity (quality is essentially the mean signal pixel intensity divided by the standard deviation of the background intensity), the target area (a function of the percentage of the expected spot area occupied by signal above background), the background flatness (a measure of the extent to which the local background exceeds the mean background across a large portion of the array), and the signal intensity consistency (a measure of the variability of the signal pixel intensity). All four of the quality measures are transformed to lie between 0 and 1, and the overall quality of the spot is taken to be the minimum of the four measures. This approach allows analysts to exclude measurements at individual spots from further analysis if the quality is too low.

Other approaches to assessing spot quality have been proposed. Wang and colleagues have also developed a quantification software package that provides multiple measures of spot quality [Wang *et al.*, 2001]. Brown and colleagues use the "spot ratio variability", which divides the pixel-by-pixel standard deviation of the ratio by the mean ratio [Brown *et al.*, 2001]. Tran and colleagues examine the difference between the mean signal intensity and median signal intensity of the pixels within a spot, requiring that they differ by less than 10% at good spots [Tran *et al.*, 2002].

4. QUALITY CONTROL OF INDIVIDUAL SLIDES

After a hybridized slide has been scanned and quantified, several methods can be applied to assess the quality of the experiment. The immediate goal of these tests is usually to decide rapidly if the experiment needs to be repeated.

4.1 View the images

The simplest test to describe is also the hardest to quantify or to automate. Before starting the statistical analysis, we look at the scanned images of the microarray to ensure that there are no obvious gross anomalies. (This step is sometimes referred to as "cortical filtering"; in order to pass the test, the data must be successfully processed by a human cortex.) We visualize the images in MATLAB (The MathWorks, Inc., Natick MA), which allows us to alter the color map. Most people do a poor job of discriminating between different shades of gray, and they are even less good at distinguishing colors that shade from black into red or green. False color images prepared in MATLAB can give a good idea of the quality of the spots and of any strange behavior in the background.

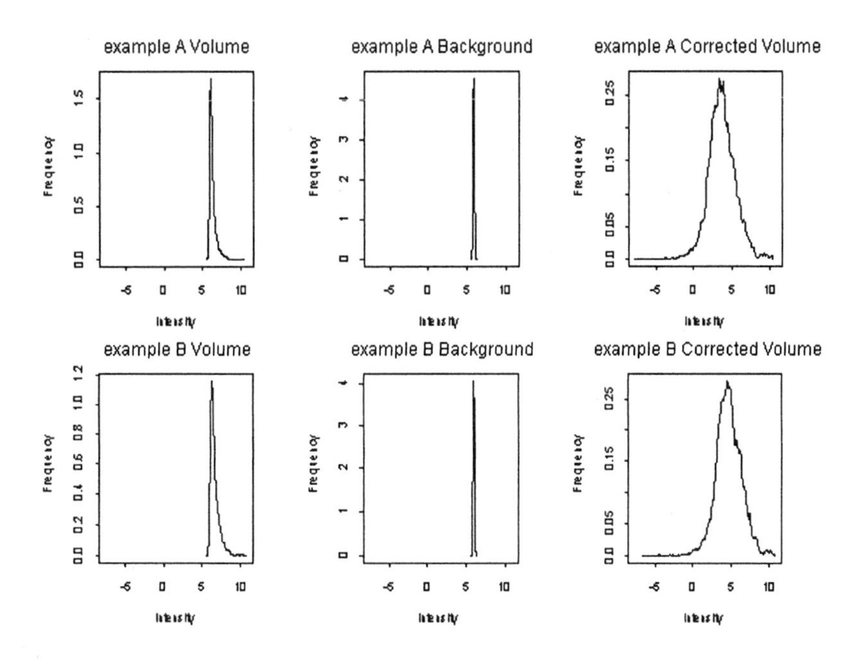

Figure 2. Plots of the estimated probability density of signal intensity (left panels), background intensity (center panels) and background-corrected signal intensity (right panels) from both the red (top row) and green (bottom row) channels.

4.2 Density estimates

After deciding that the images are adequate, we load the quantification data into S-Plus (Insightful Corp., Seattle WA). We compute kernel density estimates of the distributions of the logarithm of the raw volume intensity estimates and the raw background estimates for each channel (Figure 2). The example here shows a fairly typical distribution pattern. The volume has a high peak with a long right tail. The background has a tighter peak located at about the same point as the volume density, without the extended tail. The background-corrected volume is much more spread out. We interpret these distributions as telling us that a large number of spots are near or below the background levels; the spots that truly show significant gene expression are the ones in the long tail at the right.

Any qualitative change in these density plots is taken as an indication of a potential problem. Bimodal peaks in the background plots, for example, usually indicate that large regions are subject to excessively high background noise. They may also indicate that a portion of the grid was improperly aligned during quantification, causing the signal to be counted as background over a large subsection of the array. Obvious differences in the shape of the distributions from the red and green channels typically indicate a problem (poor quality RNA or faulty labeling) with one of the samples.

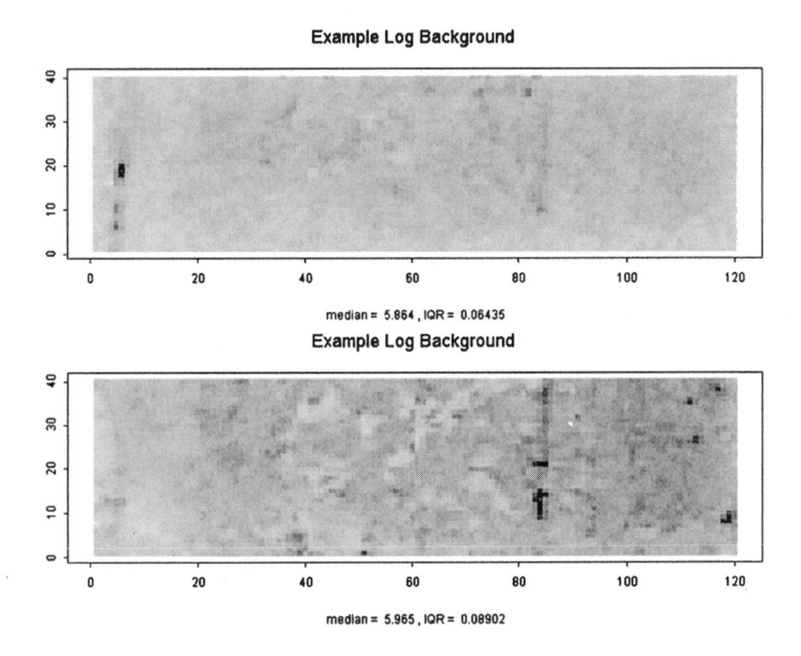

Figure 3. Spatial plots of the logarithm of background intensity in red and green channels.

4.3 Topological plots of background

We also look at "cartoon" images of the logarithm of the estimated background intensity at each spot (Figure 3). The log transformation allows us to see more detail at lower intensity levels, where the background is more important. In this example, the red channel background is plotted in the top figure and shows few significant features; the most visible glitch consists of a few spots near the left end. As usually happens, the background in the green channel is both slightly higher on average (as reflected in the estimate of the median) and more variable. In this case, the background is substantially higher on the right third of the green channel.

4.4 Signal-to-noise ratio

One of the most effective measures for the overall quality of a microarray experiment is the percentage of spots with adequate signal-to-noise ratio (S/N). The precise definition of "adequate" in this setting may vary from platform to platform. On Affymetrix microarrays, for example, the model used in their Microarray Analysis Suite version 5.0 provides detection p-values along with "present" or "absent" calls for each gene. We typically find that around 40–50% of the genes are present in a successful hybridization on an Affymetrix microarray.

On our own two-color fluorescent microarrays, we typically require good spots to have S/N > 2. We have found, for successful hybridizations, that the percentage of spots with S/N > 2 is typically in the range of 30% to 60%. We see a similar range of percentages regardless of the tissue of origin for the RNA sample. We have seen extreme cases where as few as 5% or as many as 95% of the spots pass the signal-to-noise threshold; both extremes are indications that something went wrong during the conduct of the experiment.

5. QUALITY CONTROL WITH REPLICATE SPOTS

The microarrays produced in our Genomics Core have the virtue that every gene of interest is spotted on the array in duplicate. The value of printing replicate spots to obtain more accurate expression measurements has been described previously [Ramakrishnan *et al.*, 2002]. Because of the replications, we can estimate measurement variability within the microarray. For each channel, we make scatter plots that compare the first and second member of each pair of replicates (Figure 4, left). More useful, however, are the Bland-Altman plots obtained by rotating the scatter plots by 45 degrees (Figure 4, right). To construct a Bland-Altman plot, we graph the difference between the log intensity of the replicates (whose absolute value is an estimate of the standard deviation) as a function of the mean log intensity. In the typical plots shown here, there is a much wider spread (a "fan" or a "fishtail") at the left end of the plot (low intensity) and much tighter reproducibility at the right end (high intensity). We can fit a smooth curve to this graph that estimates the variability (in the form of the absolute difference) as a function of the mean intensity. In Figure 4, we have plotted confidence bands at plus and minus three times those smooth curves. Replicates whose difference lies outside these confidence bands are intrinsically less reliable than replicates whose difference lies within the bands [Baggerly *et al.*, 2000; Tseng *et al.*, 2001; Loos *et al.*, 2001].

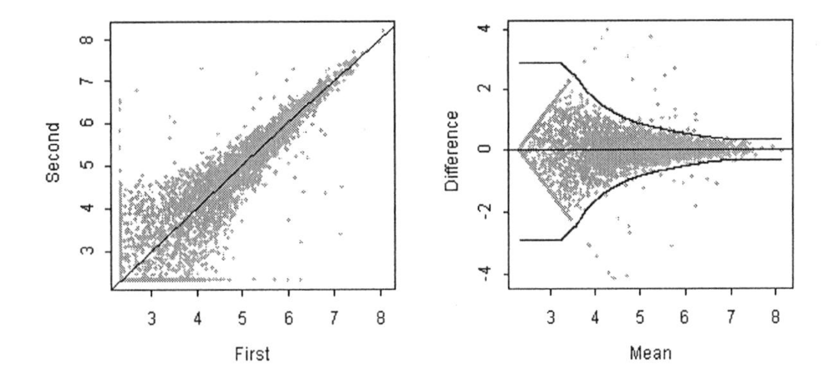

Figure 4. Quality control using replicate spots. (Left) Scatter plots of the replicate log intensities of genes in the red channel. (Right) Same plots after rotating by 45 degrees. Curves represent three times a smooth (loess) estimate of the absolute difference.

6. QUALITY CONTROL USING REPLICATE EXPERIMENTS

All microarray experiments must include some degree of replication if we are to assign any meaningful statistical significance to the results. In the previous section, we showed how some simple plots can automatically detect spots where replicated spots give divergent measurements. The same method can be used with replicated experiments. This method works with purely technological replicates (separate hybridizations of the same labeled RNA, separately labeling the same RNA, or separate RNA extractions from the same mixture of cells) or with biological replicates (in which RNA samples are independently obtained under similar biological conditions [Novak *et al.*, 2002; Raffelsberger *et al.*, 2002]. When replication occurs later in the process, of course, one expects the scatter plot of the replicates to follow the identity line more closely.

Scatter plots of replicate experiments can detect two other common problems. The first problem can occur when merging data into a spreadsheet for analysis. It is all too easy to sort data from different arrays in different orders, which causes the data rows to be mismatched. If the entire array is affected, then a plot of the replicates shows an uncorrelated cloud of points instead of the expected band along the identity line (Figure 5). Mismatching a subset of the rows produces a plot that combines part of the expected band with a smaller cloud. This phenomenon can also occur when a few of the subgrids on one array were misaligned during quantification (Figure 5).

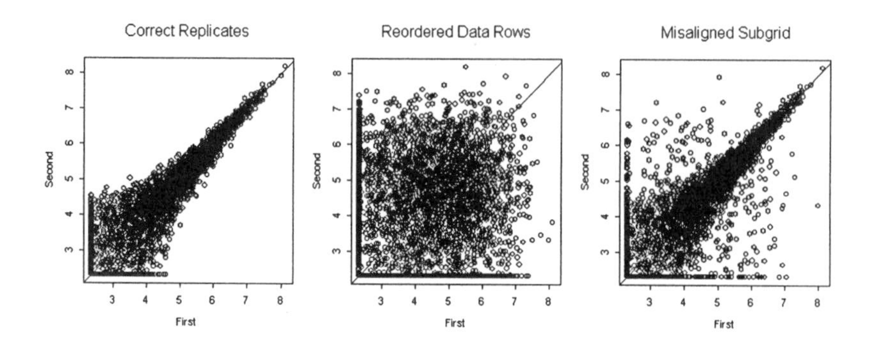

Figure 5. Scatter plots of replicate experiments. (Left) The expected scatter plot for technological or biological replicate experiments. (Center) A typical plot when the data from one replicate has been accidentally reordered. (Right) Typical plot when a small number of subgrids has been misaligned, equivalent to randomly reordering a subset of the data rows.

7. CLUSTERING FOR QUALITY CONTROL

In some array studies, the investigators make a clear distinction between class discovery and discrimination. Supervised methods (linear discriminant analysis or support vector machines) are appropriate if the goal of the study is discrimination or diagnosis. Unsupervised methods (hierarchical clustering, principal components analysis, or self-organizing maps) are generally appropriate if the goal of the study is class discovery. In some published studies, unsupervised methods have been used regardless of the goal. The attitude seems to have been: "because our unsupervised method rediscovered known classes, our microarrays are as good at pattern recognition as a diagnostic pathologist." In our opinion, such an attitude is misguided.

Nevertheless, unsupervised classification methods do have an appropriate place in the analysis of microarray data even when the correct classification of the experimental samples is known. The avowed purpose of unsupervised methods is to uncover any structure inherent in the data. If they recover known structure, then that result is hardly surprising. Because the statistical and mathematical properties of these methods have not been fully investigated in the context of microarray data, neither can one draw strong conclusions from the recovery of known structure. As we have pointed out, many things can go wrong during the process of collecting microarray data. If an unsupervised classification method yields deviations from the known structure, these deviations can be quite revealing of problems in data quality.

We have successfully used clustering methods to identify many of the problems described in earlier sections of this paper. When the study includes technological replicates of each experiment, we expect the replicates to be nearest neighbors (or at least close neighbors), as portrayed, for instance, in a dendrogram based on a hierarchical cluster analysis. We have seen instances where replicates failed to be neighbors; in every case, the cause of the failure could be traced to a problem with the data. The problems detected using this method have included, among others:

- individual arrays where the data rows had been reordered,
- misalignment of the entire grid or of subgrids,
- analytical errors where "red – green" differences had accidentally been combined with "green – red" differences,
- differences between microarrays produced in different print lots, and
- differences in the dynamic range of signals from different arrays.

Plotting the samples using the first two or three components from a principal components analysis (PCA) can also reveal similar anomalies. A striking application of this method can be found elsewhere in this volume [Stivers *et al.*, 2003]. By applying PCA to the individual channels of the Project Normal microarray experiments, we found that the reference channels could be separated based on the tissue of origin of the sample hybridized to the other channel. This finding led us to look more closely at the data. In turn, this led to the discovery that the UniGene annotations in data files from experiments using liver RNA did not agree with the UniGene annotations in data files from experiments using kidney RNA.

8. PROCESS CONTROL

All the methods discussed so far deal with quality control rather than process control. The distinction is simple. Quality control is based primarily on inspections. Its goal is to identify components of low quality and reject them. We have described methods to inspect RNA samples and microarray slides before hybridization. We have described various methods to inspect the scanned image of a microarray experiment and determine the quality of the hybridization. We have also looked at methods to inspect the quality of individual spots. In this context, the use of clustering methods plays the role of a “final inspection” before an entire collection of microarray experiments is released to the “consumer” of the data.

Process control, on the other hand, is focused not on inspections but on monitoring. Its goal is to detect changes in the process in real time in order both to maintain and to improve the overall quality [Ryan, 1989]. In a typical academic microarray facility, it is hard to see how to apply process control to the manufacture of microarrays. Such facilities do not, in general, run a continuous production line to print microarrays. Instead, batches of arrays are printed sporadically, as needed. Acceptance testing for the quality control of batches of printed arrays seems more appropriate. Many academic microarray facilities, however, do perform hybridizations on a full-time basis, day in and day out. If the goal of a microarray core facility is the production of successful microarray hybridizations, then there is an important role for the statistical methods of process control to maintain and improve the quality of those hybridizations.

Model and colleagues have described a way to apply statistical process control to microarray experiments [Model *et al.*, 2002]. Their method starts with a robust version of PCA to identify individual microarray experiments as outliers. The robust PCA is used to compute a version of Hotelling's T^2 statistic, which is a measure of the distance from the array to the mean (over all arrays) in a d-dimensional principal component space. This statistic should have a chi-squared distribution, with d degrees of freedom. Individual outlier arrays are removed, and multivariate statistical process control is implemented using a T^2 control chart [Ryan, 1989, chapter 9].

In Model's method, PCA is performed on the log ratios between the channels of a two-color microarray experiment. Their intended application is to large-scale microarray projects that hybridize several hundred microarrays using biologically similar samples. As they point out, the experimental design must include randomization or block randomization in order to ensure that a change in the biology of the samples over time is not misinterpreted as an out-of-control process.

As it stands, the method cannot readily be applied to a microarray core facility that is performing hybridizations for a series of small to medium size projects, with each project looking at different sample types for different investigators. However, if all investigators use the same common reference material in one channel as part of their experimental design, then a modified method can be used. Principal components analysis can be performed on the intensities measured in the common reference channel, and the T^2 control chart can be constructed using these data. In this way, the operation of the core facility can be monitored in real time, independently of the nature of the biological samples being used by individual investigators.

9. CONCLUSION

The complex process of conducting a series of microarray experiments affords numerous opportunities for errors to occur. In this article, we have begun to describe a comprehensive approach to quality control and quality improvement. We have discussed methods that can be employed at different steps in the process. We recommended establishing standard protocols and solid database foundations, to try to avoid common errors. We have described methods that can be used during array manufacture, before hybridization, and after hybridization. We have described methods for assessing the quality of individual spots, of individual slides, and of complete sets of microarray experiments. Finally, we have described how ideas from process control can be used to monitor the performance of a microarray core facility over time. Quality control measures need to be applied at all stages of the process. We believe that the methods discussed here form a solid foundation on which to develop more robust and reliable ways to carry out microarray studies.

REFERENCES

Baggerly KA, Coombes KR, Hess KR, Stivers DN, Abruzzo LV, Zhang W. Identifying differentially expressed genes in cDNA microarray experiments. J Comput Biol. 2001; 8:639-59

Battaglia C, Salani G, Consolandi C, Bernardi LR, De Bellis G. Analysis of DNA microarrays by non-destructive fluorescent staining using SYBR green II. Biotechniques. 2000; 29:78-81.

Brazma A, Hingamp P, Quackenbush J, Sherlock G, Spellman P, Stoeckert C, Aach J, Ansorge W, Ball CA, Causton HC, Gaasterland T, Glenisson P, Holstege FC, Kim IF, Markowitz V, Matese JC, Parkinson H, Robinson A, Sarkans U, Schulze-Kremer S, Stewart J, Taylor R, Vilo J, Vingron M. Minimum information about a microarray experiment (MIAME)-toward standards for microarray data. Nat Genet. 2001; 29:365-71.

Brazma A, Sarkans U, Robinson A, Vilo J, Vingron M, Hoheisel J, Fellenberg K. Microarray data representation, annotation and storage. Adv Biochem Eng Biotechnol. 2002; 77:113-139.

Brown CS, Goodwin PC, Sorger PK. Image metrics in the statistical analysis of DNA microarray data. Proc Natl Acad Sci U S A. 2001; 98:8944-9.

Chen Y, Kamat V, Dougherty ER, Bittner ML, Meltzer PS, Trent JM. Ratio statistics of gene expression levels and applications to microarray data analysis. Bioinformatics. 2002; 18:1207-15.

Dunmire V, Wu C, Symmans WF, Zhang W. Increased yield of total RNA from fine-needle aspirates for use in expression microarray analysis. Biotechniques. 2002; 33:890-896.

Hegde P, Qi R, Abernathy K, Gay C, Dharap S, Gaspard R, Hughes JE, Snesrud E, Lee N, Quackenbush J. A concise guide to cDNA microarray analysis. Biotechniques. 2000; 29:548-50, 552-4, 556 passim.

Hsiao LL, Jensen RV, Yoshida T, Clark KE, Blumenstock JE, Gullans SR. Correcting for signal saturation errors in the analysis of microarray data. Biotechniques. 2002; 32:330-2, 334, 336.

Hu L, Cogdell DE, Jia YJ, Hamilton SR, Zhang W. Monitoring of cDNA microarray with common primer target and hybridization specificity with selected targets. Biotechniques. 2002; 32:528, 530-2, 534.

Loos A, Glanemann C, Willis LB, O'Brien XM, Lessard PA, Gerstmeir R, Guillouet S, Sinskey AJ. Development and validation of corynebacterium DNA microarrays. Appl Environ Microbiol. 2001; 67:2310-8.

Model F, König T, Piepenbrock C, Adorján P. Statistical process control for large scale microarray experiments. Bioinformatics. 2002: 18 Suppl 1; S155-S163.

Novak JP, Sladek R, Hudson TJ. Characterization of variability in large-scale gene expression data: implications for study design. Genomics. 2002; 79:104-113.

Pritchard CC, Hsu L, Delrow J, Nelson PS. Project normal: defining normal variance in mouse gene expression. Proc Natl Acad U S A. 2002; 98:13266-13271.

Raffelsberger W, Dembele D, Neubauer M, Gottardis M, Gronemeyer H. Quality indicators increase the reliability of microarray data. Genomics. 2002; 80:385-394.

Ramakrishnan R, Dorris D, Lublinsky A, Nguyen A, Domanus M, Prokhorova A, Gieser L, Touma E, Lockner R, Tata M, Zhu X, Patterson M, Shippy R, Sendera TJ, Mazumder A. An assessment of Motorola CodeLink microarray performance for gene expression profiling applications. Nucleic Acids Res. 2002; 30:e30.

Ryan T. Statistical Methods for Quality Improvement. John Wiley and Sons, New York, 1989.

Shearstone JR, Allaire NE, Getman ME, Perrin S. Nondestructive quality control for microarray production. Biotechniques. 2002; 32:1051-2, 1054, 1056-7.

Stivers DN, Wang J, Rosner GL, Coombes KR. Organ-specific differences in gene expression and UniGene annotations describing source material. In: Methods of Microarray Data Analyis III (Lin SM, Johnson KF, eds), Kluwer Academic Publishers, Boston, 2003.

Taylor E, Cogdell D, Coombes K, Hu L, Ramdas L, Tabor A, Hamilton S, Zhang W. Sequence verification as quality-control step for production of cDNA microarrays. Biotechniques. 2001; 31:62-5.

Tran PH, Peiffer DA, Shin Y, Meek LM, Brody JP, Cho KW. Microarray optimizations: increasing spot accuracy and automated identification of true microarray signals. Nucleic Acids Res. 2002; 30:e54.

Tseng GC, Oh MK, Rohlin L, Liao JC, Wong WH. Issues in cDNA microarray analysis: quality filtering, channel normalization, models of variations and assessment of gene effects. Nucleic Acids Res. 2001; 29:2549-57.

Wang X, Ghosh S, Guo SW. Quantitative quality control in microarray image processing and data acquisition. Nucleic Acids Res. 2001 Aug 1;29(15):E75-5.

Weil MR, Macatee T, Garner HR. Toward a universal standard: comparing two methods for standardizing spotted microarray data. Biotechniques. 2002; 32:1310-4.

Yue H, Eastman PS, Wang BB, Minor J, Doctolero MH, Nuttall RL, Stack R, Becker JW, Montgomery JR, Vainer M, Johnston R. An evaluation of the performance of cDNA microarrays for detecting changes in global mRNA expression. Nucleic Acids Res. 2001; 29:E41-1.

3

OUTLIERS IN MICROARRAY DATA ANALYSIS

Ronald K. Pearson, Gregory E. Gonye, and James S. Schwaber
The Daniel Baugh Institute for Functional Genomics and Computational Biology, Department of Pathology, Anatomy, and Cell Biology, Thomas Jefferson University, Philadelphia, PA,

Abstract: This paper presents a broad survey of analysis methods that have been found useful in the detection and treatment of outliers, or "anomalous data points," that often arise in large datasets. One source of these data anomalies is *method outliers* like "gross measurement errors" that are of no inherent biological interest, but a second source is *biological outliers,* which may be of considerable interest, for example as diagnostics for diseases or other important phenomena. This paper presents a simple illustration of the influence of outliers on microarray data analysis, considers the practical aspects of outlier detection, and briefly describes a number of results from the literature on outlier-resistant analysis methods. Most of the references discussed in this survey are not drawn from the microarray data analysis literature, but rather from the more general robust statistics literature, to provide a useful "cross-fertilization" of ideas.

Key words: outliers, robust statistics, microarray data analysis, data quality

1. INTRODUCTION

The labeling error in the CAMDA normal mouse dataset uncovered by Stivers *et al.* [2002] emphasizes the importance of detecting and appropriately treating anomalies in microarray data analysis: 1,932 inconsistently labeled genes were discovered in the three datasets made available to the CAMDA 2002 participants. Undetected, such a labeling inconsistency would lead to inaccurate analysis results and incorrect biological interpretations. Conversely, Tukey [1977, p. 49] describes a very different example of an important data anomaly: Lord Rayleigh's analysis of 15 weight measurements made on a standard volume of a gas that was believed to be nitrogen. Seven of these samples were obtained by removing oxygen from air, while the other eight were obtained from nitrogen-containing compounds, and small but consistent differences were observed between these two groups of measurements. Subsequent investigation led to Lord Rayleigh's discovery of argon, the first known inert gas, for which he was awarded the Nobel Prize in Physics in 1904. Both of these examples illustrate the importance of detecting atypical observations in datasets.

Typical microarray datasets are large enough that the search for data anomalies must be largely automated, detecting potential anomalies and bringing them to our attention for further scrutiny. This paper presents a broad survey of analysis methods that have been found useful in the detection and treatment of *outliers* or "anomalous data points" that often arise in large datasets. The definition of the term "outlier" adopted here is that offered by Barnett and Lewis [1984, p. 4]:

> We shall define an outlier in a set of data to be an observation (or subset of observations) which appears to be inconsistent with the remainder of that set of data.

It is important to distinguish between *outliers*, which represent unexpectedly large deviations in a dataset, and *noise*, which represents low-level irregularity in a dataset. In particular, most standard analysis procedures are designed to accommodate the unavoidable presence of noise, but these procedures may fail completely in the presence of a few outliers, as the example discussed in Section 3 below demonstrates. It is also important to note that the question of how outliers should be treated depends in part on how we choose to interpret them. In particular, if we assume they arise from *gross measurement errors*, we are led to questions of how best to clean the dataset so that these data anomalies do not corrupt our analysis results.

Conversely, if the outliers represent atypical but biologically significant results, we are led in the direction of discovery science. As a specific example, the search for anomalous gene expression levels in tumor samples vs. normal tissue samples may be viewed as a search for outliers with possible diagnostic applications. Here, we will adopt the terms *method outliers*, to refer to gross measurement errors and other outliers of no inherent biological interest, and *biological outliers* to refer to atypical results that may have potential diagnostic or other biological utility. In either case, it is important to note that while the problem of *detecting* outliers in a dataset can be approached using mathematical methods, the problem of *interpreting* these outliers is *not* a mathematical one, but depends on our knowledge of the biology and the measurement system.

2. THE INFLUENCE OF OUTLIERS

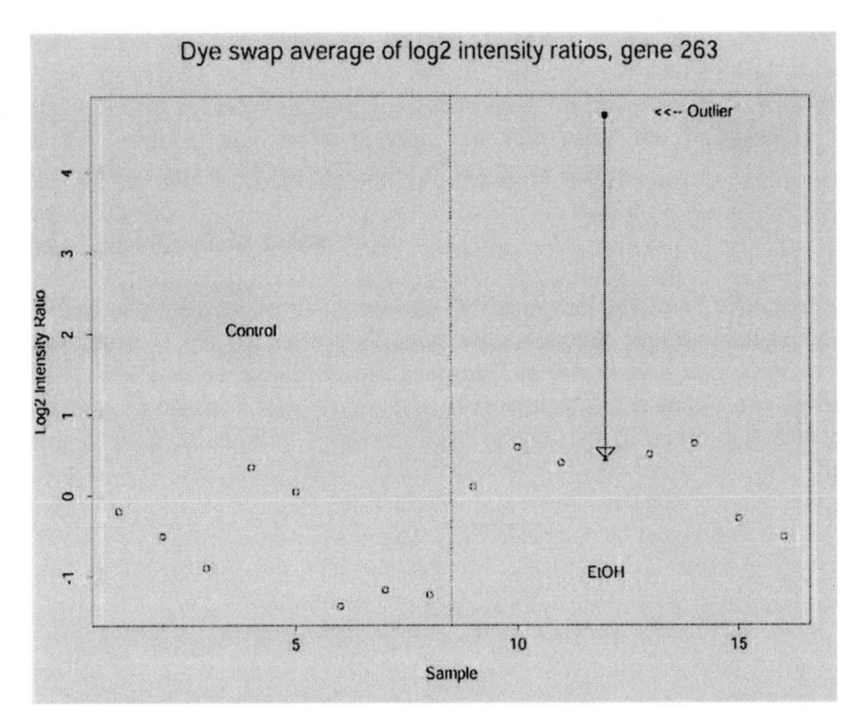

Figure 1. An outlier and its median replacement: The open circles represent 15 typical $\log_2$ intensity ratios for replicated measurements of a single gene in both control and ethanol-treated animals. The solid circle represents a single outlier and the solid triangle at the point of the downward arrow represents its median replacement value.

Figure 1 shows a sequence of cDNA microarray measurements obtained for a single gene in a series of experiments designed to measure the expression changes in rat brainstem samples in response to prolonged ethanol exposure [Milano *et al.*, 2002]. The 8 values to the left of the dashed line in this figure represent the averages of Cy3 and Cy5 dye labelings of $\log_2$ intensity ratios for individual animal mRNA samples against a common control mix reference prepared from four control animals, and the 8 values to the right of the dashed line are the corresponding measurements for the ethanol-treated animals. Each experiment was replicated twice for each dye labeling and each of the four animals, giving a total of 16 control and 16 treatment samples; averaging the dye swaps to reduce dye bias yields the 16 numbers shown in Figure 1. The key point here is that one of these measurements (observation no. 12) appears as a visually obvious outlier.

Once we have found an outlier, we are faced with the question of what to do with it, and as noted earlier, the answer to this question must depend on whether we regard the outlier as a method outlier or a biological outlier. If we regard it as a method outlier, a reasonable strategy is to replace it with a more representative value, computed from the rest of the available data. In the example shown in Figure 1, a useful approach is to replace the outlying ethanol-treated value with the *median* of the 8 ethanol-treated values. In general, given a sequence of N real numbers, the median is computed by first sorting them into ascending order and, if N is an odd number, the median is equal to the unique middle element in this sorted list. If N is an even number, the median is defined as the average of the two middle values. Since the median is extremely insensitive to outliers, it is often a reasonable replacement value for outliers. For this example, the effect of this median replacement strategy is shown in Figure 1 with the arrow from the original outlying value to its median replacement. Clearly, the median replacement is more consistent with the rest of the ethanol-treated results than the original value is. Although the source of this particular data anomaly is unknown, the large difference between this value and both its replicate, obtained from the same mRNA sample, and the pairs of samples from the other ethanol-treated animals strongly suggests a method outlier.

Under the common working assumption that most genes are not responsive to ethanol exposure, the search for genes that exhibit a strong ethanol response represents a search for biological outliers. A standard tool for detecting these anomalous genes is a volcano plot like the one shown in Figure 2. This plot shows the difference in mean $\log_2$ intensity ratios between 8 ethanol-treated and the 8 control animals for 100 genes, plotted against the p-value computed from a t-test of the null hypothesis that these means are the same. The horizontal dashed lines in this figure represent differences in $\log_2$ intensity ratio of ± 1 unit, corresponding to 2-fold up- or down-regulation of the indicated gene. The two vertical dashed lines correspond to the 1% and 5% thresholds for the t-test. Biological outliers should appear as genes with small p-values, but the results we obtain can be profoundly influenced by the presence of method outliers. This point is illustrated by the four special cases represented by arrows, connecting open circles with solid circles. These genes are like the one shown in Figure 1 where the individual expression values contain visually obvious outliers: the open circles show the results obtained from the unmodified data values, while the solid circles show the results obtained when these outlying values are replaced by the appropriate median values. The key point here is that these data modifications cause large shifts in both the relative expression changes and the computed p-values. Further, note that these changes cause the affected points to be moved into positions that are generally more consistent with the pattern of the volcano plot generated by the other 96 genes. Also, since the individual data anomalies in these four cases are anomalously large values, they cause the computed expression changes to be overestimated; hence, outlier replacement reduces these expression change values, but the computed p-values can either increase or decrease.

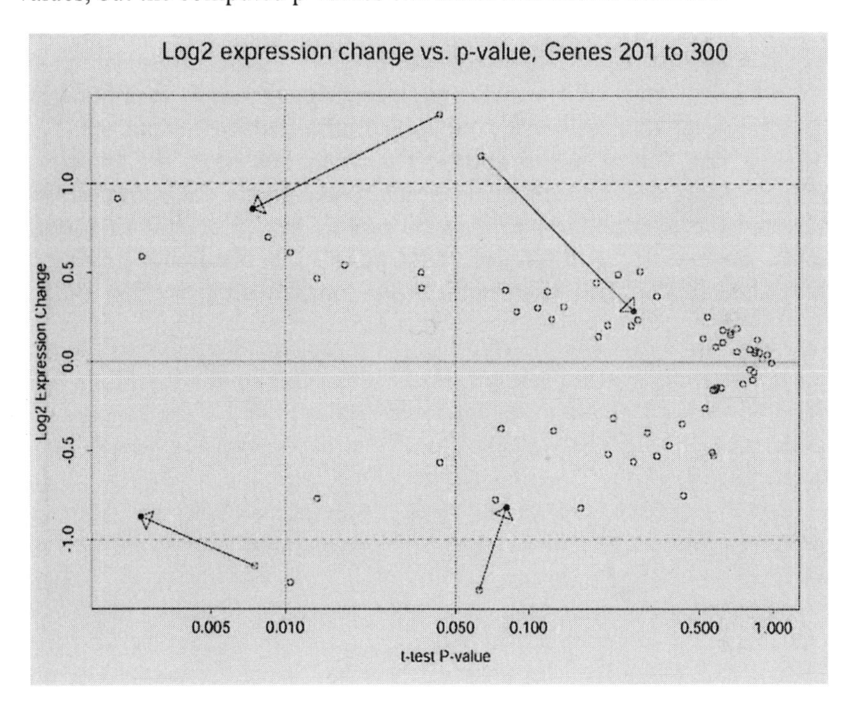

Figure 2. Volcano plot: Mean $\log_2$ expression ratio changes between treatment and control results are plotted against the p-values computed from a standard t-test. The four arrows show the changes in these results caused by correcting method outliers in the original data.

3. DETECTION OF OUTLIERS

The example shown in Figure 1 provides a clear, graphic illustration of the nature of outliers in microarray data, but as noted at the beginning of this paper, visual inspection is not a practical basis for outlier detection in large datasets. Instead, we need an automated outlier detection procedure that classifies data points into "nominal" and "anomalous." Without question, the most popular procedure for detecting outliers is the "3σ edit rule," an idea that goes back at least to the end of the nineteenth century [Wright, 1884]. This procedure is based on the working assumption that the nominal data values form an independent, identically distributed (i.i.d.) sequence of Gaussian random variables. Under this assumption, the probability of observing a value lying more than three standard deviations from the mean is only about 0.3%. Hence, given a set of N data observations, the 3σ-edit rule consists of the following steps. First, compute the mean μ and the standard deviation σ from the data sequence. Then, for each individual data point, compute the absolute distance it lies from the mean μ. If this distance is larger than 3σ, declare the point an outlier. This procedure has also been called the *extreme studentized deviation (ESD) identifier* [Davies and Gather, 1993]. Despite its popularity, this procedure tends to be ineffective in practice *even if the Gaussian assumption for the nominal data is reasonable*, due to the phenomenon of *masking*. Specifically, the presence of outliers in the dataset tends to introduce biases in both the mean and the standard deviation. As a consequence, the ESD identifier frequently fails to identify outliers that are present in a dataset. This may be viewed as a particularly ironic example of the influence of outliers: their presence causes an apparently well-founded outlier detection procedure to fail.

This point is illustrated in the left-hand plot in Figure 3, which shows a sequence of eight ostensibly identical gene expression measurements from a sequence of microarray experiments. It is clear from visual inspection that the observed response from sample number 4 is inconsistent with the range of variation seen in the other seven samples, but this observation is not detected as an outlier by the ESD identifier. The dashed lines in the figure represent the mean value and the upper and lower outlier detection limits.

An alternative to the ESD identifier is the *Hampel identifier* [Davies and Gather, 1993], which replaces the mean with the median and the standard deviation with the *median absolute deviation (MAD)* in the procedure outlined above. The MAD scale estimator is defined as follows. First, compute the median of the data sequence and then form the sequence of differences between each data value and this median. Next, compute the absolute values of these differences, giving a sequence of nonnegative numbers that measure the distance each data point lies from the median reference value. Then, compute the median of this sequence of absolute values, giving a measure of how far data points *typically* lie from the median reference value. Finally, multiplying this value by 1.4826 yields an unbiased estimate of the standard deviation for Gaussian nominal data:

$$S = 1.4826 \text{ median } \{ \mid x_k - \text{median } \{ x_k \} \mid \}$$

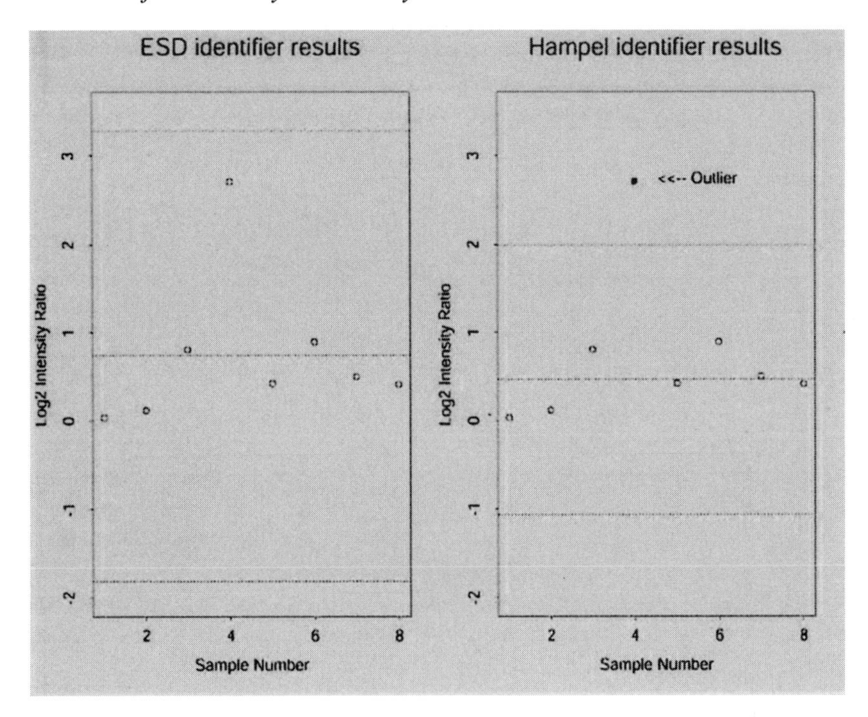

Figure 3. ESD vs. Hampel outlier detection procedures: The left-hand plot shows the mean and ESD outlier detection limits as horizontal dashed lines for a sequence of eight data points that includes a visually obvious outlier. The right-hand plot shows the same data points with horizontal dashed lines indicating the median and the Hampel outlier detection limits.

Huber [1981] has described the MAD scale estimate as, "perhaps the most useful ancillary estimate of scale," and it has been used in the microarray data analysis community for robust data normalization procedures [Yang *et al.*, 2002]. Replacing the mean and standard deviation with the median and MAD scale estimate in the 3σ-edit rule leads to an outlier identification procedure that is much less prone to masking effects. The right-hand panel in Figure 3 shows the same 8 data points as the left-hand panel, but the dashed lines in this plot correspond to the median and the upper and lower outlier detection limits for the Hampel identifier. Sample number 4 is clearly detected as an outlier by this procedure.

Other outlier-resistant outlier detection procedures may be developed by replacing the mean and standard deviation in the 3σ-edit rule with other alternatives. One alternative to the mean is the *trimmed mean,* discussed further in Section 4, and various alternatives to the standard deviation are possible [Rousseeuw and Croux, 1993], including the *interquartile distance (IQD)*, defined as the distance between the *upper quartile* (the smallest data value larger than or equal to 75% of the total data values), and the *lower quartile* (the smallest data value larger than or equal to 25% of the total data values). As in the case of the MAD scale estimator just discussed, to obtain an unbiased estimate of the standard deviation σ for Gaussian data, it is necessary to divide the IQD by 1.35 [Venables and Ripley, 2002, p. 122]. Although it is not as outlier-resistant as the MAD scale estimate [Huber, 1981, p. 107], the IQD is commonly used in detecting outliers in boxplots. As a specific example, the default for the **boxplot** command in the *S-plus* statistical analysis package [Statsci, 2000] is to declare any point lying more than 1.5 IQD's beyond the upper or lower quartiles as outliers and mark them individually as horizontal line segments on the boxplots.

An important point is that effective outlier detection procedures should tolerate some violation of the working assumptions on which they are based. It is clear from the discussion presented here that the 3σ-edit rule fails in this regard, whereas the other two outlier detection procedures described here are more tolerant of violations of this same working assumption: namely, that the nominal data values are approximately normally distributed. In fact, these procedures often work reasonably well even with strongly non-Gaussian nominal data, provided the distribution is not too markedly asymmetric. This observation is important since many useful ideas cease to be applicable to strongly skewed data distributions. For example, the mean and the median both estimate the same location parameter for any symmetric distribution like the Gaussian, but this is no longer true for asymmetric distributions like the lognormal. Although some outlier-resistant characterizations are available for asymmetrically distributed data, like the alternatives to the MAD scale estimate described by Rousseeuw and Croux [1993], the general difficulty of the situation is summarized by J.W. Tukey's comments in the summary of the Princeton Robustness Study, a thorough comparison of the small-sample performance of 68 different location estimators [Andrews *et al.*, 1972, p. 226]:

> We did a little about unsymmetrical situations, but we were not able to agree, either between or within individuals, as to the criteria to be used.

The issue of asymmetry is particularly relevant to microarray data analysis since optical intensity values exhibit markedly asymmetric distributions. Hence, even in cases where popular logarithmic data transformations do not result in a normally distributed result, they do significantly reduce distributional asymmetry, making outlier detection procedures more applicable than they would be if applied to raw intensity values.

Finally, the question of the sample size N is also particularly important in outlier detection, since performance generally becomes less reliable as N decreases. One useful observation is that, for a procedure like the Hampel identifier that can tolerate anything less than 50% contamination, outlier detection is theoretically possible for sample sizes as small as N = 3, although care is required since in any sequence containing two identical values and a third distinct value, this third value will *always* be declared an outlier by the Hampel identifier. Conversely, it is not difficult to show that this third value will *never* be declared an outlier by the ESD identifier. Overall, it is even more important to use outlier-resistant detection procedures like the Hampel identifier for small samples than for large ones.

4. A SURVEY OF SOME OTHER USEFUL IDEAS

The statistical literature on outliers and closely related topics is much too large to review in detail here. Further, many publications dealing with outliers in specific contexts describe techniques that are much more broadly applicable, specializing them through the choice of illustrative examples. As a case in point, although the applications considered are the analysis of industrial process data and laboratory physical properties data, the paper by Pearson [2001] discusses a number of generally useful techniques for discovering data anomalies, many of which are equally applicable to the analysis of microarray data (e.g., the use of quantile-quantile plots to detect outliers). The following paragraphs focus on a few useful ideas, citing references for more detailed discussions and more complete bibliographies.

Barnett and Lewis [1984] present a very readable, broad survey of the topic of outliers, including much historical detail. Topics covered include location estimation (e.g., estimation of the mean of the nominal data subsequence from the outlier-contaminated sample), scale estimation (e.g., the standard deviation of the nominal data subsequence alone), hypothesis tests like the t-test, estimation of mean vectors and correlation matrices from contaminated data samples, regression models, principal component analysis, and time-series characterizations. It is worth emphasizing that, however well they perform in the absence of outliers, many standard analysis procedures suffer badly in the presence of outliers. For example, Devlin *et al.* [1981] describes a principal components analysis example in which the first principal component is entirely due to a single outlier.

Also, these procedures depend strongly on the assumptions made both about the nominal part of the data sequence and about the outliers. For example, two outlier models considered by Barnett and Lewis are the *slippage model* and the *mixture model.* In their most common forms, slippage models assume that outliers are random variables with a different mean from the nominal data, whereas mixture models typically postulate that the outliers have a larger variance than the nominal data. While Barnett and Lewis devote more attention to the slippage model than to the mixture model, many authors adopt the contaminated normal mixture model discussed by Huber [1981]. The difference is important in practice because slippage outliers or "shift outliers" can cause difficulties for procedures based on the contaminated normal model [Woodruff and Rocke, 1994].

Barnett and Lewis also devote much discussion to procedures for detecting outliers in different situations: univariate data sequences (where outliers are necessarily extreme values), multivariate data sequences, regression models, and time-series models. In addition, they also illustrate the use of ancillary variables in confirming suspicious points as outliers. Specifically, they describe a regression problem based on measurements of auricular pressure and coronary flow in cats published by Rahman [1972]. Although the bulk of the data appears consistent with a linear relation between these variables, four observations are visually inconsistent with this relationship. Further, these four observations all occur for the *same* value of auricular pressure; subsequent examination of the original data values show that they are a sequence of *consecutive* measurements that decay monotonically to zero, further supporting the hypothesis that these four observations are anomalous. Barnett and Lewis raise the question of whether this particular cat died during the experiments.

Despite the wealth of details and references presented by Barnett and Lewis, many of the outlier detection procedures described by these authors make the historically important but extremely restrictive assumption that the number of outliers in the data sequence is very small. In particular, many of the procedures described explicitly assume that at most one or two outliers are present in the dataset. Although these procedures may be extended by iterative application, this basic assumption is highly questionable in the analysis of large datasets. In fact, the question of how many outliers should be considered in a dataset has been, and remains, quite controversial. For example, one proposal discussed by Barnett and Lewis [1984, p. 140] is to consider $k = \sqrt{N}$ as an upper limit on the number of outliers possible in a dataset of size N. The authors quote Chhikara and Feiveson [1980] as noting:

> It is reasonable to consider three potential outliers in a data set of 10 observations, but it is unrealistic to expect 30 outliers out of a data set of 100 observations. In the latter case, the outlier detection problem becomes one of discrimination between two or more classes of data.

If we apply the $\sqrt{N}$ limit to the CAMDA normal mouse dataset, containing 5304 observations, we arrive at an upper limit of 73 outliers; unfortunately, Stivers *et al.* [2002] detected 1932 anomalous observations, representing approximately 36% of the dataset. The practical importance of this point lies in the fact that many different analysis procedures are continuing to emerge from the field of robust statistics, driven by the practical problem of effectively analyzing contaminated data sequences. In the most extreme cases, estimators like the median and the MAD scale estimate can tolerate any fraction of contamination less than 50% and still give arguably reasonable characterizations of the nominal part of the data. Two prices we usually pay for this extreme outlier resistance are first, a worsening in performance relative to less outlier-resistant estimators when outliers are absent from the sample and second, an increase in computational complexity. These considerations have led Huber [1993] to argue that:

> Contaminations between 1% and 10% are sufficiently common that a blind application of procedures with a breakdown point of less than 10% must be considered outright dangerous. But it is difficult to find a rational justification for a breakdown point higher than 25% if this should result in a serious efficiency loss at the model, or in excessive computation times.

It is important to note that many "standard" data characterizations, including both the mean and the standard deviation, exhibit 0% breakdown points: a single outlier of sufficient magnitude in an arbitrarily large dataset can completely dominate the results.

An important class of methods that frequently satisfies Huber's desirability criteria quoted above is the class of *M-estimators*, which he was instrumental in developing [Huber, 1981]. Motivation for this idea comes from the fact that many "classical" data analysis methods may be recast as least-squares optimization problems. That is, we construct a simple mathematical model that predicts our observed data sequence on the basis of a small number of parameters. One simple but extremely useful data model is the *location model,* which assumes that all observed values are approximately equal to some unknown constant, μ. Associated with this data model is a sequence of *prediction errors*, equal to the difference between each individual data observation and its value as predicted by the data model. *Least squares* procedures for estimating the parameters in this data model select those parameter values that minimize the sum of squared prediction errors. These procedures have the practical advantage of often yielding simple, explicit solutions, but they almost always exhibit high outlier sensitivities. As a specific example, for the location model, the least squares solution is the mean of the data sequence.

The reason for the pronounced outlier-sensitivity of least squares procedures is that they attach very large fitting penalties to large errors. One alternative that overcomes this difficulty is *least absolute deviations (LAD)* [Bloomfield and Steiger, 1983], which minimizes the sum of absolute values of the model prediction errors instead of the sum of squared prediction errors. Historically, this method dates back to Galileo in 1632 and was studied by Laplace and others in the late eighteenth century [Dodge, 1987]. If we consider the location estimation problem, the LAD solution is the sample median. Although LAD problems generally have solutions that, like the median, are much more outlier-resistant than their least-squares counterparts, they suffer two limitations relative to least squares. First, the required computations are usually much harder and second, the optimal parameter vector is often non-unique. Huber's M-estimator approach attempts an intermediate course, minimizing a quantity that penalizes small errors quadratically like least squares but penalizes large errors linearly like LAD procedures. Huber [1981] shows that the resulting solutions are unique and may be computed by a simple iterative extension of standard weighted least squares procedures, called *iteratively reweighted least squares.*

Rousseeuw and Leroy [1987] describe a different alternative to the standard least squares optimization problem, leading to analysis procedures that exhibit the same extreme outlier-resistance as the median and the MAD scale estimate, but suffering from both of the difficulties noted earlier: performance suffers relative to alternatives like M-estimators in the absence of outliers, and the computational complexity is substantially greater. Still, the basic idea is a clever one and it has led to the development of useful computational procedures. The essential notion is to replace the *sum* of squared deviations (equivalent to minimizing the average squared deviation) with the *median* of squared deviations, leading to a class of estimators known as *least median of squares (LMS)* procedures. In practice, a better-behaved alternative is to replace the median in this prescription with the *trimmed mean*, a simple idea that is commonly used in grading tests and judging athletic contests to reduce the influence of outlying scores. Specifically, the symmetric α-trimmed mean of a sequence of N numbers is computed by first rank-ordering the numbers just as in computing the median, then eliminating the bottom and top fractions α of these values and averaging those remaining. As noted by Barnett and Lewis [1984, p. 23], this idea goes back at least to Mendeleev at the end of the nineteenth century, and it represents a simple way of interpolating between the arithmetic mean (corresponding to $\alpha = 0\%$) and the median (corresponding to the limit as $\alpha \rightarrow 50\%$). Replacing the median of the squared prediction errors with an intermediate trimmed mean leads to the class of *least trimmed squares (LTS)* estimators, which are generally better-behaved than the LMS-estimators that exhibit non-normal limiting distributions [Rousseeuw and Leroy, 1987]. It is worth noting that both the M-estimators described earlier and the LTS-estimators just described are iterative procedures that involve either down-weighting (in the case of M-estimators) or rejecting (in the case of LTS-estimators) data points that give rise to very large fit errors.

The notions underlying LMS and LTS estimators are closely related to methods developed for the detection of *multivariate outliers*. In contrast to the univariate case where outliers are necessarily extreme values, outliers in a collection of vector-valued data objects need not exhibit extreme values with respect to any individual component of these vectors. Instead, it is enough that a particular data object be anomalous with respect to the *relationship between these components* that is exhibited by the nominal data observations, making them generally harder to detect than univariate outliers. In the context of microarray data analysis, one potential application of multivariate outlier detection procedures would be to treat expression levels for a particular gene as a vector with components corresponding to expression levels in different tissue types. Multivariate outliers in this setting would correspond to genes that violated the relationship between these expression levels seen in normal tissue samples.

One approach to the multivariate outlier detection problem uses the *minimum volume ellipsoid (MVE)*, defined as the ellipsoid of smallest volume that includes at least 50% of the data values [Rousseeuw and Leroy, 1987, p. 258]. A direct search for this ellipsoid would require examination of all size N/2 subsets for a dataset of size N, rapidly leading to computational infeasibility and motivating the development of heuristic approaches based on random subsample selection [Atkinson, 1994; Rousseeuw and Leroy, 1987; Woodruff and Rocke, 1994]. As a specific example, Atkinson [1994] used an MVE-based procedure to search for outliers in a dataset containing 8 physical property measurements for each of 86 milk samples. This procedure detected 17 outliers, only one of which exhibited extreme values with respect to the individual physical property measurements, but all of which were clearly evident as outliers when the appropriate scatterplots of one component versus another were examined. Also, note that the number of outliers detected in this case is about twice the $\sqrt{N}$ upper limit discussed by Barnett and Lewis [1984] for the maximum number of outliers to consider in a dataset of size N.

Finally, although they cover a much wider range of topics beyond outliers, Venables and Ripley [2002] present brief but extremely readable and useful discussions of a number of the ideas described here, including the MAD and IQD scale estimators, M-estimators, LMS- and LTS-estimators. Another feature that makes these discussions particularly useful is that they describe specific implementations available in the *R* and *S-plus* statistics packages, both based on the *S* language developed at Bell Laboratories. The *S-plus* package is commercially available from Insightful Corporation (http://www.insightful.com) and the *R* package is an Open Source system (http://www.r-project.org) that has become reasonably popular in the microarray data analysis community.

5. SUMMARY

This paper has attempted to give a brief but broad survey of the problem of outliers, outlier detection, and outlier treatment in the analysis of microarray data. Most of the ideas discussed here have been presented in the broader context of exploratory data analysis, rather than the specific context of microarray data analysis, but the relevance of these ideas to microarrays was demonstrated through the examples discussed in Sections 2 and 3. Again, it is worth emphasizing that the data anomalies discovered by Stivers *et al.* [2002] in the analysis of the CAMDA normal mouse cDNA dataset illustrate both the relevance of the issues discussed here and the fact that anomalous data values need not occur either in isolation or in "small concentration," as often assumed. Further, mechanisms like that responsible for the CAMDA normal mouse data anomaly can be expected to become increasingly common with the increased use of automated analysis procedures that combine experimental data with data retrieved from public repositories. It is also important to bear in mind first, that the *interpretation* of outlying observations once we have found them is not a mathematical problem since it necessarily involves issues of biology and measurement methodology and second, that outliers may be of inherently greater interest than a careful characterization of the nominal data. This was arguably the case in Lord Rayleigh's discovery of argon and it is certainly the case in the search for genes with diagnostic utility that respond preferentially to various pathologies. Finally, two points are worth emphasizing: first, that many standard analysis procedures (e.g., least squares regression analysis) are extremely sensitive to the presence of outliers, and second, that one of the best techniques for uncovering outliers is to compare the results obtained by different methods (e.g., means vs. medians) or from different datasets that are expected to give similar results. Large discrepancies may indicate the presence of outliers and should probably be investigated.

ACKNOWLEDGMENTS

The authors wish to thank their colleagues Jan Hoek and Ausra Milano for providing the microarray data discussed in Sections 2 and 3, and Hester Liu for technical assistance. Funding support for this work was provided by

NIH/NIGMS project MH64459-01, NIH/NIAAA project AA13203-01, and DARPA project BAA0126.

REFERENCES

Andrews, D, Bickel, P, Hampel, F, Huber, P, Rogers, W, and Tukey, J (1972) Robust Estimates of Location. Princeton University Press.

Atkinson, A (1994) Fast very robust methods for the detection of multiple outliers. J Amer Statist Assoc 89:1329-1339.

Barnett, V and Lewis, T (1984) Outliers in Statistical Data. Wiley, 3rd ed.

Bloomfield, P and Steiger, W (1983) Least Absolute Deviations. Birkhauser.

Chhikara, R and Feiveson, A (1980) Extended critical values of extreme studentized deviate test statistics for detecting multiple outliers. Commun Statist Sim Comp 9:155-166.

Davies. L and Gather, U (1993) The identification of multiple outliers. J Amer Statist Assoc 88:782-801

Devlin, S, Gnanadesikan, R and Kettenring, J (1981) Robust estimation of dispersion matrices and principal components. J Amer Statist Assoc 77:354-362

Dodge, Y (1987) An introduction to statistical data analysis: L1-norm based. In: Dodge, Y (ed) Statisticsl Data Analysis Based on the L1-Norm and Related Methods. North-Holland.

Huber, PJ (1981) Robust Statistics. Wiley.

Huber, PJ (1993) Projection pursuit and robustness. In: Morenthaler, S, Ronchetti, E, and Stahel, W (eds) New Directions in Statistical Data Analysis and Robustness, pp. 139-146. Birkhauser.

Milano, A, Gonye, G, Schwaber, J, Pearson, R, and Hoek, J (2002) "Gene Expression in Rat Liver, Heart, and Brain During Chronic Alcoholism," in preparation.

Pearson, RK (2001) Exploring process data. J Process Control 11:179-194.

Rahman, N (1972) Practical Exercises in Probability and Statistics. Griffin.

Rousseeuw, P and Croux, C (1993) Alternatives to the median absolute deviation. J Amer Statist Assoc 88:1273-1283.

Rousseeuw, P and Leroy, A (1987) Robust Regression and Outlier Detection. Wiley.

Statsci (2000) S-PLUS 6 User's Guide. Mathsoft Inc.

Stivers DN, Wang J, Rosner GL, Coombes KR. Organ-specific differences in gene expression and UniGene annotations describing source material. In: Methods of Microarray Data Analyis III (Lin SM, Johnson KF, eds), Kluwer Academic Publishers, Boston, 2003.

Tukey, JW (1977) Exploratory Data Analysis. Addison-Wesley.

Venables, WN and Ripley, BD (2002) Modern Applied Statistics with S. Springer-Verlag.

Woodruff, D and Rocke, D (1994) Computable robust estimation of multivariate location and shape in high dimension using compound estimators. J Amer Statist Assoc 89:888-896.

Wright, T (1884) A Treatise on the Adjustment of Observations by the Method of Least Squares. Van Nostrand.

Yang, Y, Dudoit, S, Luu, P, Lin, D, Peng, V, Ngai, J, and Speed, T (2002) Normalization for cDNA microarray data: a robust composite method addressing single and multiple slide systematic variation. Nucleic Acids Research 30.

SECTION II

BEST PRESENTATION AWARD

4

ORGAN-SPECIFIC DIFFERENCES IN GENE EXPRESSION AND UNIGENE ANNOTATIONS DESCRIBING SOURCE MATERIAL

David N. Stivers, Jing Wang, Gary L. Rosner, and Kevin R. Coombes
University of Texas M.D. Anderson Cancer Center, Houston, TX.

Abstract: For this paper, we analyzed data collected for Project Normal, which assessed the normal variation of gene expression in three distinct mouse organs. Our approach uncovered two difficulties with the original analysis. First, normalization using the loess method obscures a difference in the distribution of gene expression measurements between pure organ samples and the reference mixture. Second, and more importantly, it appears that the link between spot location and gene annotations was broken during data processing. We used principal components analysis to confirm that reordering the data matrices based on spot locations produced more consistent data from the reference channels. In order to recover the true gene annotations, we then introduced a model that explicitly accounted for organ-specific differences in gene expression and related those differences to UniGene annotations that describe mRNA sources. Using this model, we were able to determine which set of annotations was correct.

Key words: Microarray, normalization, UniGene, permutation test, biological reference material, abundant genes.

1. INTRODUCTION

A central principle of modern biology is that differences in gene expression are directly responsible for differences in phenotype. For instance, kidney cells and liver cells from the same organism share exactly the same complement of genetic material; their differences result from the fact that kidney cells express one set of genes and liver cells express another set. Pritchard and colleagues performed a series of microarray experiments that allows us to test this principle directly [Pritchard *et al.*, 2001]. They took six genetically identical C57BL6 male mice of the same age, sacrificed them, and removed their kidneys, livers, and testes. RNA was extracted from each organ of each mouse and used for multiple microarray experiments. The analysis of the data by the original authors focused on identifying genes whose normal variation within an organ was especially large; in this article, we focus on identifying genes expressed in one organ but not in the others.

2. REFERENCE AND NORMALIZATION

In the original analysis of this data set, Pritchard and colleagues used the pin-by-pin loess normalization method introduced by Dudoit and colleagues [Dudoit *et al.*, 2002]. In the present study, we investigated the appropriateness of this transformation. To start, we used a simple normalization method. We computed the signal in each channel by subtracting local background values from measured foreground values. We divided all signals within a channel by the 75th percentile and then multiplied by 10. (We use the 75th percentile in order to be certain that we are normalizing to the value of a gene that is truly expressed.) Normalized signals less than the arbitrary threshold value of 0.5 were replaced by the threshold, after which we computed the logarithm base two. As a standard diagnostic tool, we prepared A-M graphs for each microarray [Altman and Bland, 1983]. An A-M graph plots the log ratio (difference between channels) as a function of the abundance (average log signal across both channels). Figure 1 contains A-M plots for all four microarray experiments performed using RNA from the liver of the second mouse. Plots for other organ-mouse combinations are similar to these (data not shown).

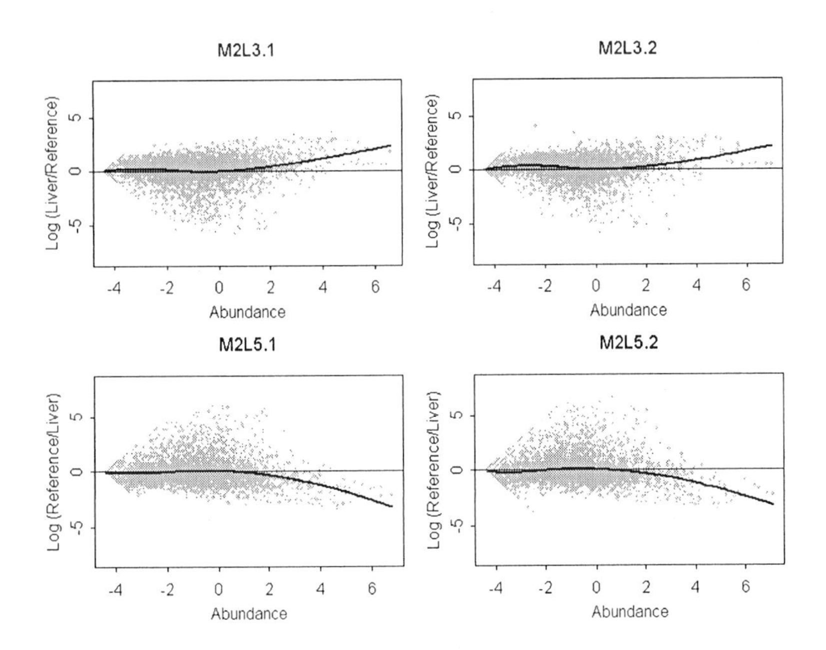

Figure 1. Plots of the log ratio (difference) as a function of the abundance (mean log intensity) for four experiments using RNA from the liver of mouse 2.

We interpret these graphs differently than do the original authors. We agree that each A-M graph shows a significant nonlinear trend reflecting a difference between the two channels. However, dyes, arrays, or other technological factors are not the cause of this nonlinearity. We base this claim on the observation that the shape of the nonlinearity is consistent from array to array, but the direction changes depending on whether the reference sample is used in the red channel (top two plots) or in the green channel (bottom two plots). We believe that the nonlinearity results from a true biological difference in the distributions of gene intensities between the reference channel, which was an equal mixture of RNA from three different organs, and the experimental channel, which contained RNA from a single organ. We explain why these distributions should be different in the following section.

2.1 Theoretical justification for the existence of different distributions of gene intensity

The reference material was the same on all 72 microarrays in this set of experiments. The investigators made the reference material by mixing equal amounts of total RNA from all 18 mouse-organs used as experimental samples. Thus, we see that the intensities of the genes in Figure 1 arise from a mixture of three distributions:

1. Genes expressed at similar levels in all three organs should be present at the normal range in the reference mixture and should give log ratios centered at zero when we compare an experimental channel from one organ to the reference mixture.
2. Genes expressed in liver, for example, but expressed at low levels or unexpressed in kidney and testis should be present in a compressed range of values in the mixture because they are diluted to one-third. Thus, they should give log ratios near log(3).
3. Genes unexpressed in liver, for example, but expressed in kidney or in testis should also be present at a compressed range of values in the reference mixture and should yield log ratios smaller than log(1/3).

If the number of genes in case 2 and 3 is substantial, then the signals present in the reference mixture should be compressed when compared to the signals from pure organs. This compression of the signal range would lead directly to the kinds of nonlinearities seen in Figure 1. We performed simulations based on a simple model incorporating this kind of mixture distribution. We modeled the observed signal, Y, using the equation

$$Y = (S + C)e^{\eta} + \varepsilon$$

where S represents gene-specific hybridization, C represents the contribution from cross-hybridization, and $\eta \sim N(0, 0.25)$ and $\varepsilon \sim N(0, 10)$ represent normally distributed multiplicative and additive noise. Because genes with high levels of expression are rare compared to genes with low, we modeled the distribution of the true signal S using an exponential distribution with mean 1000. We assumed that cross-hybridization was the same for all organs but contributed different amounts to the expression of individual genes because of sequence-dependent differences. So, we modeled C using an exponential distribution with mean 10. We then assumed that 2000 genes were expressed in common in all three organs and 1000 genes were unique to each organ. The organ-specific genes, then, had a true signal mean expression of 1000/3 in the simulated reference mixture and a true signal mean of either 1000 (if present) or 0 (if absent) in a single organ.. The resulting A-M plots are similar to the plots from the actual data (Figure 2).

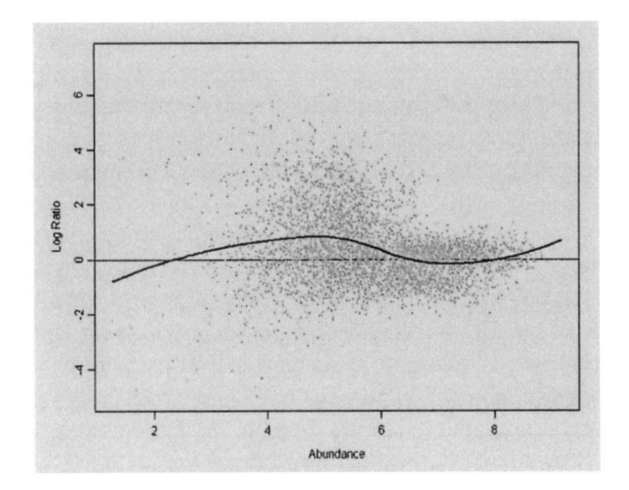

Figure 2. An A-M plot of data simulated from a model with unbalanced differentially expressed genes.

Performing a loess smooth to remove this nonlinearity, then, would remove significant biological structure that is present in the data as a direct consequence of the design of the reference material. This procedure would make it more difficult to determine which genes are specifically expressed in which organs. Consequently, we decided not to use a loess normalization method on this data, and worked with data normalized to the 75th percentile in each channel. We also realized that we would have to work with the channels separately in order to get a better idea how to proceed.

3. DATA CLEANUP

The data downloaded from the web site consisted of three files of tab-separated-values, one file for each organ. After confirming that the clone identifiers and UniGene cluster numbers were consistent in these files, we performed a principal components analysis on the log transformed normalized channel data. We plotted the first two principal components against each other (Figure 3A). We note several features of this graph. First, the experimental channels from kidney, liver, and testis can easily be distinguished from one another. Unfortunately, the reference channels (which repeatedly use the same mixture of material) can also be distinguished based on the tissue that was used in the other channel of the same experiment. Twenty of the reference channels from testis experiments are close to the reference channels from liver experiments; the other four (circled in the graph) are close to the reference channels from kidney experiments. Moreover, the distance between the reference channel experiments is about the same as the distance between experiments from different organs.

Because we expected the reference channels to be closer together, we conjectured that an error had occurred in assembling the data files. We returned to the data files and looked at the annotations that described the position of spots for the same clone ID on the microarray (block, row and column). These positions did not match in the three data files. We reordered the data matrices to align data rows by spot position and performed another principal components analysis (Figure 3B). In this case, the reference channels from experiments using liver or kidney were close together, along with four of the reference channels from testis experiments. Twenty other reference channels from testis experiments were farther away.

At this point we realized that, using either method to merge the data, the reference samples formed two groups. When the data is merged based on UniGene annotations, one group of reference samples consists of all samples from liver experiments and 20 samples from testis experiments; the other group consists of all samples from kidney experiments and four samples from testis experiments. When the data was joined based on spot location, the samples from all liver and kidney experiments and from 4 testis experiments formed the first group, and the remaining 20 samples from testis experiments formed the second group. Thus, we further conjectured that the error in data processing had occurred in the middle of the testis experiments. We reordered the data rows of the samples in a manner consistent with this conjecture and performed a third principal components analysis (Figure 3C). In this graph, all three sets of reference experiments are clustered near the center of the graph, showing much less difference within this group than between the groups of experiments using pure organs. We believe that this graph provides strong evidence in favor of our conjecture.

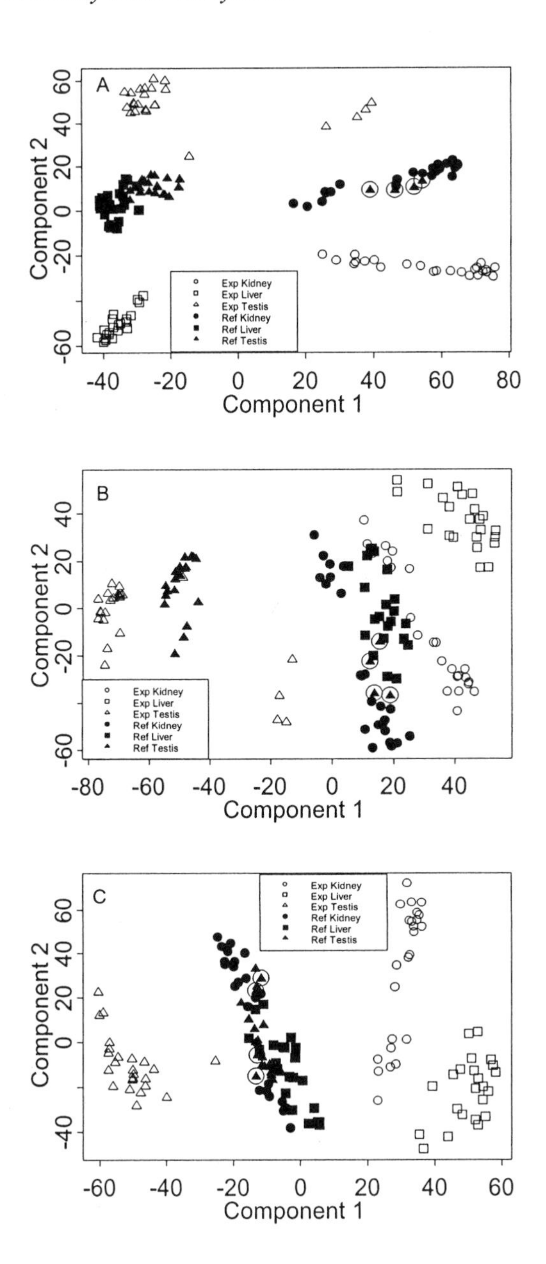

Figure 3. Plots of the first two principal components, joined by (A) Unigene number, (B) position (block, row, and column), and (C) as described in text.

4. AMBIGUOUS ANNOTATIONS

The error that we discovered in ordering the data rows has serious consequences for a biological interpretation of the data from this experiment. After reordering the data to align block, row, and column annotations for the kidney and liver experiments, the UniGene annotations at some spots no longer match in the two files.

We say that a spot has an ambiguous UniGene annotation when the files do not match; otherwise, we say it has a consistent UniGene annotation. We found that 1932 (approximately 36%) of the 5304 non-control spots on the microarray have ambiguous UniGene annotations. Thus, we can no longer reliably identify the genes at these spots. Instead, we must decide which of two equally plausible sets of UniGene annotations is correct. In light of this difficulty, we focused the remainder of our analysis on the following questions. Can we determine which spots are expressed specifically in each of the three organs? Can we combine this information with other biological information about the clone sets to determine which of the two possible UniGene annotations of the disordered spots is more likely to be correct?

4.1 UniGene Expression Data

UniGene includes a list of the organs that are the source of mRNAs included in each cluster. We downloaded this information for all genes on the microarray (ftp://ftp.ncbi.nih.gov/repository/UniGene/MM.data.Z). For each UniGene number associated with a spot on the array, we recorded a three-dimensional binary indicator vector **I** = (K,L,T). The value K = 1 was used to indicate that at least one mRNA source for the UniGene cluster was derived from kidney, while K = 0 was used to indicate the absence of any kidney-derived mRNA in the UniGene cluster. We will refer to genes for which K = 1 as "abundant" in kidney and to genes for which K = 0 as "rare" in kidney. Similar interpretations were used for liver (L) and testis (T). The number of genes in each abundance class is shown in Table 1, which also includes these numbers for genes with consistent or ambiguous notations in this data set.

Table 1. Size of abundance classes of genes, by quality of spot UniGene annotation.

Abundance	All UniGene	Consistent	Ambiguous
None	409	237	172
Kidney	129	76	53
Liver	284	169	115
Testis	372	231	141
Kidney, Liver	126	69	57
Kidney, Testis	226	146	80

Abundance	All UniGene	Consistent	Ambiguous
Liver, Testis	960	609	351
Kidney, Liver, Testis	2798	1835	963
Total	5304	3372	1932

It is important to note that the number of spots with ambiguous abundance class annotations is smaller than the number of spots with ambiguous UniGene annotations. Of the 1932 spots with ambiguous UniGene annotations, only 1310 had ambiguous abundance class annotations. Simply, this occurs because different UniGene clusters can be expressed abundantly in the same way across the three organs included in this study. In other words, there are spots that are identified as different UniGene clusters in the two files, but the abundance vector **I** associated with the two UniGene clusters is the same.

4.2 Modeling the Expression of Rare and Abundant Genes

Our basic idea is that the measured log intensities of genes that are abundant in a given tissue should be higher, on average, than the measured log intensities of genes that are rare. Moreover, we should be able to combine the UniGene annotations with the observed expression values to determine which set of UniGene annotations is correct. To illustrate this idea, consider the gene villin, which belongs to UniGene cluster Mm.4010. The list of cDNA sources for the clones that are members of this UniGene cluster includes kidney but does not include liver or testis. We would, therefore, expect the microarray experiments to detect significantly higher levels of expression of this gene in kidney compared to the other two organs. One set of UniGene annotations (which came with the kidney data) claims that villin is located in Block 2, Column 17, Row 5. The other set of annotations (which came with the liver data) says that villin is located in Block 4, Column 17, Row 5. A box plot of the measured gene expressions, by organ, at these spots, strongly suggests that villin is more likely to be located at the first spot (Figure 4).

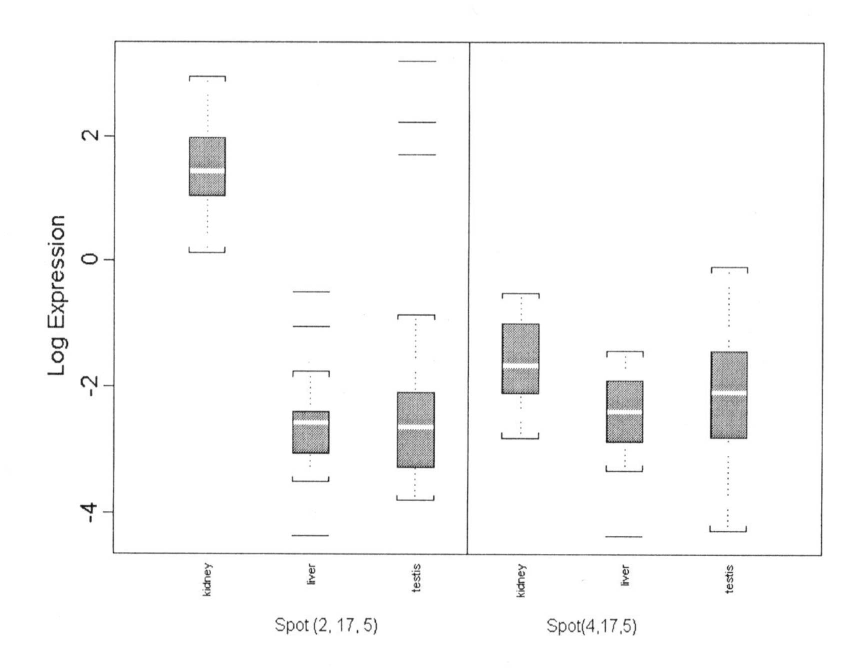

Figure 4. Box plots, by organ, of the logarithm of gene expression measured at two spots in microarray experiments. Each set of annotations identifies one of these spots as villin, which is abundant in kidney but not in liver or testis.

Let $\mathbf{Y}$ = (k, l, t) denote a triple of log intensity expression measurements from three organs of one mouse at one microarray spot. Let $\mathbf{I}$ = (K,L,T) be a binary indicator vector as above. We propose the model

$$\mathbf{Y_I} \sim N_3(\mu_\mathbf{I}, \Sigma_\mathbf{I}),$$

where N_3 denotes a three-dimensional multivariate normal distribution, $\mu_\mathbf{I} = (\mu_K, \mu_L, \mu_T)$ is a vector of mean expression that depends on whether a gene is abundant or rare in each tissue type, and $\Sigma_\mathbf{I}$ is a 3x3 covariance matrix.

Table 2. Mean log expression estimated from the model. Values predicted to be higher are indicated with boldface.

Abundance	μ_K	μ_L	μ_T
None	2.027	2.129	2.012
Kidney	**2.445**	1.880	1.822
Liver	1.911	**2.909**	1.743
Testis	1.734	1.809	**2.872**
Kidney, Liver	**3.282**	**3.051**	1.961

Abundance	μ_K	μ_L	μ_T
Kidney, Testis	**2.410**	2.129	**2.521**
Liver, Testis	2.438	**2.563**	**2.526**
Kidney, Liver, Testis	**3.202**	**3.121**	**2.958**

In order to apply this model, we must find a natural way to associate separate experimental measurements on the three organs into a three-dimensional vector. To accomplish this goal, we first averaged the pairs of replicate experiments labeling mRNA from the same organ of the same mouse with the same dye color. Specifying the mouse (of six) and the color (of two) allows us to identify a unique triple of experiments performed on the three organs from that mouse. Using these measurement triples from spots with consistent, unambiguous notations, we were able to estimate the parameters in our model. The estimated parameters for the mean log expression in each abundance class are displayed in Table 2. The estimated means are consistent with the idea that motivated the model; namely, genes belonging to a UniGene cluster whose members were derived from a specific organ should have greater mean expression when measured in that organ. The table also suggests that there is a common mean for abundant genes (and another for rare genes) independent of tissue, but we have not pursued this idea.

4.3 Distinguishing between sets of annotations

The next step is to apply the model in a manner that can distinguish between competing sets of UniGene annotations. Using the estimated parameters, we computed the likelihood of seeing the observed data at each ambiguous spot for each of the eight potential indicator vectors describing an abundance class. Thus, given a complete set of prospective abundance class annotations for the ambiguous spots, we could compute the log-likelihood of that set of annotations. We found that the log-likelihood of observing all the actual data at the spots with ambiguous abundance annotations, given the prospective annotations coming from the file of kidney data, was –52241. Given the prospective annotations coming from the file of liver data, the log-likelihood was –60183. This suggests quite strongly that the kidney file was more likely to contain the correct set of gene annotations. However, we also found that the maximum likelihood set of annotations had a log-likelihood of –33491. Thus, we were not yet certain how significant this result was.

4.4 Statistical Significance

The two competing sets of annotations for the ambiguous spots are related in the sense that one is a permutation of the other. That is, we believe that the original error arose from reordering the gene annotations before attaching them to the measured expression data. If that is indeed the case, then the log-likelihood coming from the liver file annotations should be equivalent to a log-likelihood arising from a random permutation of the ambiguous annotations, whereas the log-likelihood arising from the kidney file annotations should be much more extreme. To assess this possibility, we performed a permutation test. We randomly generated 100 different permutations of the ambiguous annotations and computed the log-likelihood associated with each one. This allowed us to assign an empirical p-value to the log-likelihoods computed in the previous section. The empirical p-values were $p < 0.01$ for the log-likelihood computed from the kidney file annotations, and $p = 0.57$ for the log-likelihood computed from the liver file annotations. In other words, none of the random permutations had larger log-likelihoods than did the kidney file annotations, and 57 of the permutations had larger log-likelihoods than did the liver file annotations. Thus we concluded, quite confidently, that the correct gene annotations came from the kidney file, and that an error had been made when combining the gene annotations with the data from the liver experiments.

5. DISCUSSION

One of the greatest challenges in dealing with microarray data is keeping track of the auxiliary information that surrounds the gene expression measurements extracted from scanned images of microarray experiments. Unlike the sequence data collected for the Human Genome Project, microarray data is highly structured. Any number of factors, including sample sources, sample preparation protocols, hybridization conditions, and microarray lot numbers can critically affect our ability to interpret the results of a set of microarray experiments. The most critical factor, however, is an adequate description of the geometric and biological design of the microarray: which genes were spotted where? Without this information, gigabytes of gene expression measurements can quickly become worthless.

Several large-scale efforts are underway to formalize the way microarray data are stored and transported [Brazma *et al.*, 2001; Brazma *et al.*, 2002; Edgar *et el.*, 2002; Spellman *et al.*, 2002]. These efforts have generated database designs, XML type definitions, and elaborate object-oriented models. We heartily support these efforts and encourage institutions and individuals to participate in the drafting and implementation of standards.

At the same time, we recognize that most microarray data sets presently extant have neither been stored nor transported under ideal conditions. The data sets from Project Normal are a case in point. The data sets posted to the web site were not primary data. Primary data are obtained by scanning one microarray at a time. The data used here were assembled into packages, probably manually using an *ad hoc* database, spreadsheet, or perl script. Under these conditions, it is remarkably easy for the row order to be changed accidentally; for example, by sorting the rows based on the values in an arbitrarily chosen column. Because the state of the art in microarray data storage and transport allows such mistakes to happen, we feel that it is extremely important to have tools that can help recover the structure as long as part of it remains intact. The model introduced in this paper makes a substantial contribution to solving that problem.

We also believe that this model has the potential to provide useful information about the relative abundance of specific genes in various organs. The maximum likelihood estimates for abundance class annotations were much more extreme than the likelihoods coming from the UniGene annotations in the file of kidney data. This observation suggests that we can use this model to update the abundance class indicators for individual genes. In other words, we can treat the indicators that we constructed from the UniGene annotations as prior information, in the Bayesian sense, about the relative abundance or rarity of the expression of a gene in various tissues. We should then be able to use the observed microarray data to update our information about abundance. In this way, the notion of abundance would be gradually divorced from the UniGene definition that we have introduced in this work. In this way, we would be able to integrate the gene expression measurements from the microarray experiments with the information contained in existing databases to improve our understanding of the biological differences between different organs.

ACKNOWLEDGMENTS

This work was partially supported by the Tobacco Settlement Funds as appropriated by the Texas State Legislature, and by a generous donation from the Michael and Betty Kadoorie Foundation.

REFERENCES

Altman DG, Bland JM. Measurement in medicine: the analysis of method comparison studies. The Statistician 1983; 32:307–317.

Brazma A, Hingamp P, Quackenbush J, Sherlock G, Spellman P, Stoeckert C, Aach J, Ansorge W, Ball CA, Causton HC, Gaasterland T, Glenisson P, Holstege FC, Kim IF, Markowitz V, Matese JC, Parkinson H, Robinson A, Sarkans U, Schulze-Kremer S, Stewart J, Taylor R, Vilo J, Vingron M. Minimum information about a microarray experiment (MIAME)-toward standards for microarray data. Nat Genet. 2001; 29:365–371.

Brazma A, Sarkans U, Robinson A, Vilo J, Vingron M, Hoheisel J, Fellenberg K. Microarray data representation, annotation and storage. Adv Biochem Eng Biotechnol. 2002; 77:113–139.

Dudoit S, Yang YH, Callow MJ, Speed TP. Statistical methods for identifying differentially expressed genes in replicated cDNA microarray experiments. Statistica Sinica 2002; 12:111–139.

Edgar R, Domrachev M, Lash AE. Gene expression omnibus: NCBI gene expression and hybridization array data repository. Nucleic Acids Res. 2002; 30:207–210.

Pritchard CC, Hsu L, Delrow J, Nelson PS. Project normal: defining normal variance in mouse gene expression. Proc Natl Acad Sci U S A. 2001; 98:13266–13271.

Spellman PT, Miller M, Stewart J, Troup C, Sarkans U, Chervitz S, Bernhart D, Sherlock G, Ball C, Lepage M, Swiatek M, Marks W, Goncalves J, Markel S, Iordan D, Shojatalab M, Pizarro A, White J, Hubley R, Deutsch E, Senger M, Aronow BJ, Robinson A, Bassett D, Stoeckert CJ Jr, Brazma A. Design and implementation of microarray gene expression markup language (MAGE-ML). Genome Biol. 2002; 3:RESEARCH0046.

SECTION III

ANALYZING IMAGES

5

CHARACTERIZATION, MODELING, AND SIMULATION OF MOUSE MICROARRAY DATA

David S. Lalush
Bioinformatics Research Center, North Carolina State University, Raleigh, NC

Abstract: We developed methods for characterizing a set of microarray images and for subsequently simulating microarray images with statistical properties similar to those in the original set. Characterization involved measuring properties of individual spots and performing analysis of variance to determine the relative contributions of individual pins used for printing and individual slides to the variation observed in spot physical properties, slide background properties, and intensity of individual genes. Slide backgrounds and individual spot nonuniformities were modeled as 2D causal Markov random fields, and parameters for these were derived from the set of real images. The results of the characterization were then used to generate realistic replicates of the original dataset that can be used for evaluating microarray data processing and analysis techniques. We demonstrated the process on a set of microarray images derived from a mouse kidney experiment. The characterization of these images showed that slides from two of the six mice have significantly different spot properties from the rest. Simulated images from the set are shown to realistically model most properties of the slides, save for large handling defects. We concluded that characterization should be an important part of any microarray experiment to maintain quality control, and that realistic simulations of microarray images can be produced using these methods.

Key words: Image analysis, Quality control

1. INTRODUCTION

Microarray technology has revolutionized genetics research since its advent [Schena, *et al.*, 1995]. The ability to simultaneously interrogate thousands of genes has opened up great opportunities to study gene function, gene regulation, and genetic variation. With this revolution, we have seen great interest in methods for analyzing the volume of data produced in microarray experiments [Kerr and Churchill, 2001, Kerr, *et al.*, 2001, Tseng, *et al.*, 2001, Wolfinger, *et al.*, 2001, Yang, *et al.*, 2002]. Unfortunately, we have not had the ability to evaluate objectively the analysis methods available to determine if they are truly giving us the answers they should.

The problem of evaluation stems from two aspects of the microarray data available to date. First, microarray experiments are expensive, so replications generally have been limited to small numbers of slides, too few to run statistically significant cross-validation experiments of data analysis techniques. Second, even if there were sufficient numbers of replicates, we have the additional problem that the true gene expression of the sample is not known, nor are many of the co-variational relationships between the genes. This is, of course, because the point of the experiment is to estimate these expression levels and relationships.

So, to objectively evaluate microarray data analysis techniques, we require a large number of replications from a set of samples whose expression properties are well characterized. We could obtain such a set with a properly designed laboratory experiment. While expensive, such a set would give us a basis from which to judge some of our data analysis techniques. Unfortunately, to judge others, we might have to produce another such experiment, or many others.

We propose a less expensive and more flexible route. Simulation of microarray datasets can give us virtually all of the replicates that we want at the cost of computer time. It also allows us to set the parameters of the experiment and to consider "what if" scenarios simply by altering the parameters of the simulation. The truth is known, since we have created it.

To be effective, however, simulation must be rooted in reality as much as possible. Thus, effective characterization and modeling of real microarray data is important to the process. In simulating, it is not so important that we match the exact properties of a real data set, since that is not really possible anyway. Rather, it is important that our simulation approximate the magnitude and properties of real variation. Then, by adjusting the parameters of the simulation, we can evaluate data analysis techniques under a variety of conditions, and compare their results to the true conditions used to generate the replicates. At that point, we will know which data analysis techniques are most robust.

Microarray *numerical* data has often been simulated to evaluate data analysis techniques [Cui, *et al.*, 2002, Dobbin and Simon, 2002, Lonnstedt and Speed, 2002, Szabo, *et al.*, 2002]. We, however, take a different approach in trying to simulate the data in its most raw form, at the level of the scanned image. Realistic simulation of microarray images will allow us more flexibility in modeling and manipulating individual sources of variation. Of course, it also adds greater complexity to the simulation process and will require detailed characterization of specific components of the microarray system.

In this paper, we detail some of the methods we use to characterize, model, and simulate the properties of microarray image data. We demonstrate our approach on a set of 24 two-color cDNA slides derived from the kidneys of six normal mice, part of a project designed to estimate normal variation in gene expression among mice [Pritchard, *et al.*, 2001]. In the following, we first explain some of the characterization and modeling methods used. Then, we show selected results from the characterization data as well as example images from our simulation set.

2. METHODS

2.1 Microarray System Model

Figure 1 shows the basic layout of the model used for our simulation. The system is modeled based on individual properties of the sample, the slide, and the pins used to print the slides. The distribution of cDNA in individual spots is primarily dependent on properties of the pins. The degree

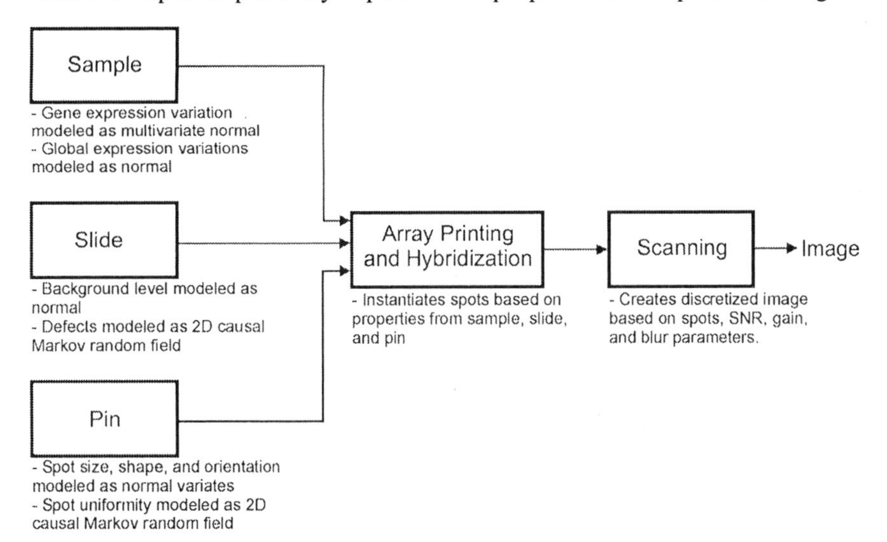

Figure 1: Summary of the structure used in the simulation process. Properties of the sample, slide, and pin are used to create a simulated spot. Properties of the slide are used to generate the background. The scanning process determines the final image.

of expression for an individual gene is dependent on a multivariate normal model that accounts for covariation between genes, but it is also dependent on an overall expression level for the given sample. Background properties are modeled entirely as dependent on properties of the individual slides.

In simulation, a slide image is generated by instantiating a collection of individual spots. Each spot is created based on random sampling from distributions governed by the parameters for its associated pin, its gene, and the slide. After creating all spots according to the defined layout for the slide, background level and slide defects are instantiated from slide parameters. The scanning process, represented by parameters governing signal-to-noise ratio and photomultiplier tube gain, creates the final image by setting intensity levels, creating random pixel noise, and discretizing the pixel values.

2.2 Characterization Methods

To create a simulation that is representative of an actual dataset, it is necessary to characterize the actual dataset in terms of the many parameters that govern the simulation model. In the following subsections, we detail the measurements and processing methods used to derive these parameters.

2.2.1 Spot Measurements

The first level of characterization is to perform measurements on the spots. We developed a program that automatically finds the spot targets, outlines them, and takes several measurements on the geometry of the spot. This program operates on the sum of the Cy3 and Cy5 images. Our spot-finding method began with finding a regularly spaced grid for each pin group within which to search for each spot. For each pin group, our program summed in the x- and y-directions to create a 1D signal for each, computed the derivative of each signal, and found zero crossings at regular intervals. The approximate spacing between spots was used as a parameter. In cases where a peak was not detected due to low signal or defects, the a priori spacing was used. In some cases, the automatic detection failed to find the starting location for a pin group, so the start location was identified manually. This method is dependent on the spot grids being aligned with the x- and y-pixel axes of the image; in some cases it was necessary to manually rotate the images into this orientation. The approximate spot grids determined from this process were used to define individual boxes in which each spot was segmented in the next step.

Within a given box, the spot was determined by starting with a 5-pixel radius circular region in the center of the box, known to be smaller than the actual spot size. Square background regions were defined in the four corners of the box, each one-sixteenth the size of the complete box. We determined the mean and standard deviation of pixels within each region. For the four background boxes, if one was significantly different than the rest, it was discarded as a possible defect. Remaining pixels in the box were scanned. Those found to be four standard deviations above the background were associated with the spot region. A five-pixel morphological "open" operation [Jain, 1989] was then applied to smooth the boundaries of the spot region and remove individual pixels. The spots were also refined by casting sixteen rays, evenly spaced over 360 degrees, from the center to measure radial distance, with regions showing ray lengths greater than two standard deviations from radial mean being corrected to the mean. At this point, if a gene was not strongly expressed, its spot size remained at the five-pixel radius.

After the spot regions were determined, each spot was measured for several quantities:

- *Radius*, computed from the average ray length;
- *Eccentricity*, intended to measure circularity of the spots [Jain, 1989], computed as the ratio of the longest ray length to the average ray length;
- *Area*, the number of pixels in the spot region;
- *Mean and standard deviation of spot pixels,* for each channel, symbolized below as μ_{signal} and σ_{signal};
- *Mean and standard deviation of background pixels,* for each channel, symbolized below as μ_{bgnd} and σ_{bgnd};
- *Separability*, for each channel, computed as follows:

$$\frac{\mu_{signal} - \mu_{bgnd}}{\sqrt{\sigma^2_{signal} + \sigma^2_{bgnd}}} \qquad (1)$$

- *Spot uniformity*, for each channel, computed as the ratio of spot pixel standard deviation to the mean: $\sigma_{signal} / \mu_{signal}$;
- *Background uniformity*, for each channel, computed as the ratio of background pixel standard deviation to the mean: $\sigma_{bgnd} / \mu_{bgnd}$.

We then performed two-way analysis of variance (ANOVA) on these measures with slide as a fixed factor and pin as a random factor, similar to methods used by others for examining intensity or intensity-ratio data [Kerr, *et al.*, 2001, Wolfinger, *et al.*, 2001]. The linear model representing this process is as follows:

$$y_{ijk} = \mu + p_i + s_j + (ps)_{ij} + e_{ijk} \qquad (2)$$

where the measured spot property, y_{ijk}, for slide i, pin group j and spot k, is composed of contributions from the overall mean value, μ, the printing pin, p_i, the slide, s_j, pin-slide interactions, $(ps)_{ij}$, and the stochastic error, e_{ijk}. Pin, pin-slide interaction, and stochastic error were assumed to be zero-mean, normally-distributed random variables.

Only those spots with a separability greater than one were used in the ANOVA analysis, as they were considered to be sufficiently expressed to make the estimated spots legitimate. The ANOVA analysis isolated the degree of variation attributable to the individual pins and to individual slides.

We checked our measured mean spot intensity values against the numerical data used by the authors of the CAMDA data set [Pritchard, *et al.*, 2001] and found that they correlated very well ($r = .987$). We take this to show that our reading of the images was comparable to that used in the original paper.

2.2.2 Background Measurements

To obtain measurements of the slide background properties, we took all pixels not defined to be in spots, and computed the mean and standard deviation of these. Currently, our model assumes that slide background properties are stationary. Observation of the true images shows that this is not the case in reality, so in the future, our model will have to be expanded to account for nonstationary backgrounds.

Background defects, unusually bright spots, marks, scratches, and other features that were significantly different from the overall background level were modeled by another process detailed in the next section.

2.2.3 Uniformity and Defect Models

Nonuniformities in microarray spots are not primarily due to noise, but rather to nonuniformities in the deposition of cDNA material on the substrate. Thus, the spot nonuniformities have structural characteristics, as one can see from observing the microarray images. To model this effect, we chose to use a 2D causal Markov random field (MRF) [Besag, 1974].

The basic approach to the MRF is shown in Figure 2. For the spots, we establish two regions. A given pixel is either in the "normal" region or the "low" region. Pixels in the normal region have one level representing their intensity; pixels in the low region have another level. The probability of being in a given region is dependent on the regions associated with the four causal neighbors shown. Thus the region model can be represented by sixteen parameters: a probability of being in region 1 associated with each of the sixteen possible states of the four neighbors. Two additional parameters give the mean and standard deviation for the intensity of the low region

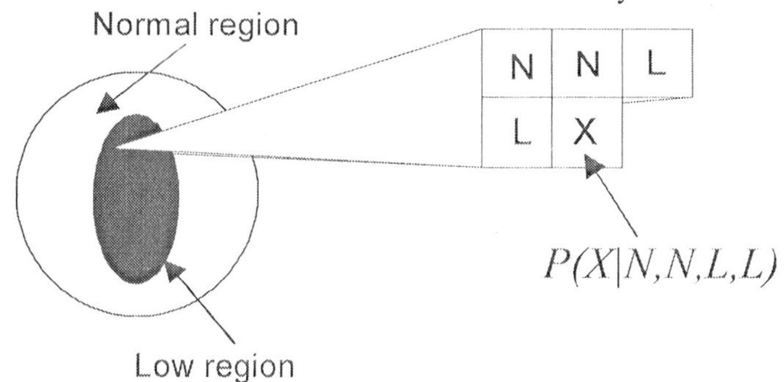

Figure 2: Illustration of the MRF model for spot nonuniformity. A spot has two regions: normal and low. Each pixel X has a probability of being normal or low conditioned on the regions of its four causal neighbors in the northwest, north, northeast, and west cardinal directions.

relative to that of the normal region, which is normalized to one. While a noncausal model would be more appropriate, the causal model was chosen for simplicity in sampling from the distribution. The choice to use four causal neighbors, rather than a more extensive neighborhood, was also made in the interest of simplicity. In the future, verification of the model will be needed to assess its appropriateness.

The two-level model was chosen to represent the property of microarray spots to have contiguous regions of low intensity, often concentrated in the center of the spot. In our model, this represents the distribution of cDNA spotting material on the slide. One could legitimately implement a three-level model with high, medium, and low levels to represent three levels of spotting material density. However, in the images, we did not observe high density regions to occur as frequently nor to have structure as significant as those of the low density regions. Thus, we chose to keep the number of parameters down and implement the two-level model.

To derive the parameters of the MRF from the images, the pixels within each spot were initially classified as either in the normal or low region. A spot was classified as "low" if its intensity was below the mean spot intensity in *both* channels. These region assignments were refined by ensuring that a pixel was only in the low region if four of its eight neighbors were in the low region as well. Then, the sixteen probabilities were computed by accumulating the number of pixels in each region for each state of the four neighbor pixels. Also, the mean and standard deviation of the low-region intensity relative to the high-region intensity was computed. The MRF parameters were computed for all of the spots in each pin group on each slide. This allowed us to perform ANOVA on the MRF parameters.

To model the tendency for low regions to concentrate in the center of a spot, we used a different non-Markov model for the center nine pixels of each spot. Here we simply derived a probability of being in the low region, independent of neighbors. From viewing our simulated images, it is not clear that the separate process is really necessary, but we have not yet studied this carefully.

Background defects were modeled in a similar 2D MRF. In this case, the regions were labeled as "normal" or "defect". A defect was taken to be a pixel more than two standard deviations above the average background for the entire image. The background pixels for each image were classified this way, and a set of MRF parameters was computed. In this case, the defect levels were characterized by their intensity relative to the background mean. The distribution of defect intensity levels was modeled with a beta random variable, with most defects having comparatively low intensity and a few having high intensity. The MRF and beta distribution parameters were determined for each slide.

2.2.4 Gene Expression Model

To capture at least some of the covariation properties between expressed genes, we modeled the set of 4700+ genes for each sample (test or reference) as a multivariate normal distribution. The expression level of each gene, which could be considered the concentration of that gene's mRNAs in the sample, is considered a vector quantity with a mean vector and a covariance matrix.

Further, we used a linear model to account for global effects from sample-to-sample that affect all genes simultaneously. This would account for differences between samples in overall concentration, dilution, contamination, and other factors. For this, we used a linear model:

$$g_{ij} = b_i + m_i \cdot a_j + \varepsilon_{ij} \tag{3}$$

In this model, g_{ij} represents the mRNA abundance of gene i in sample j. It is composed of an overall mean value, b_i, the product of a global sample contribution, a_i, and a gene-specific sensitivity to global expression variation, m_i, and a stochastic term, ε_{ij}. The overall mean and sensitivity terms are considered to be deterministic for each gene. The sample variation term is considered to be independent, zero-mean, and normally distributed. The stochastic term represents the contribution of the multivariate normal distribution.

To estimate parameters for this model, we used the measured intensities for each gene from each slide. Each set of 24 samples for each gene was regressed on the set of 24 intensities averaged over all genes in each slide. This determined for each gene a slope factor which was used for m_i. The properties of the set of 24 intensities averaged over all genes was used to estimate the properties of a_i.

We then created a new set of gene intensities by subtracting out the mean value for the gene and the regression effects. The resulting values were used to create the multivariate normal distribution. However, we did not compute the full covariance matrix. Instead, we decided to model only those correlations that were significant. We computed correlation coefficients between every pair of genes in the set over the set of 24 slides. Significant (95%) levels of correlation were computed to be .48, after using Bonferroni correction for multiple comparisons. We used an agglomerative clustering algorithm based on correlation distance to bring together groups of genes whose absolute correlations with each other were all greater than .48. We recognize that this could miss some significant correlations between genes that end up in different clusters. Starting the grouping with the highest levels of correlation first hopefully minimized such missed correlations.

A covariance matrix for each cluster was then computed and stored. By reducing the full data set into clusters, we enforce sparsity in the covariance matrix, making it easier to compute its eigenvectors and simulate.

2.3 Simulation Methods

The simulation followed the general approach shown in Figure 1. Gene expression was simulated by finding the diagonalizing transformation for each cluster with eigenvalue decomposition of its covariance matrix. Once the diagonalizing transform was found, we simulated a vector of independent normal random variables and transformed them into the proper multivariate distribution via the inverse transform.

Regions for each spot were determined by randomly sampling from the properties for its associated pin. Once a spot outline was determined, its uniformity distribution was derived by sampling from the appropriate Markov random field. This gave a distribution for each spot, which coupled with the realization of the gene expression and the gain parameter (which accounted for dye differences), resulted in base intensity levels for each pixel. Background was simulated based on sampling its associated distributions and was added into the base intensity image.

The base intensity image was blurred with a Gaussian blur of 1.5 pixels to emulate scanner resolution effects. The blur was determined solely by visual comparison between simulated and real images and should be more accurately estimated in the future. The blurred base intensity was used to generate image noise via a normal distribution on each pixel, dependent on the signal-to-noise ratio parameter that was specified by the user. Background defects were generated by sampling the 2D MRF. Finally, the image was discretized to 16 bits to emulate the TIF image format of the original images.

3. RESULTS

3.1 Characterization

The characterization and ANOVA analysis of the measured spot parameters revealed several interesting relationships:

- The variation in spot size parameters was mostly attributable to the error term. That is, most variation came from individual spots. The spot size parameters were similar for all pins (Figure 3). Looking at variation by slide, however, we observed that the morphology parameters for the last eight slides, those corresponding to mice five and six, were different from the rest, showing larger spot sizes by about 12% (Figure 4). This points to some difference in the processing of the last eight slides which we will return to in other observations below.
- Eccentricity variation was strongly associated with variations in spots

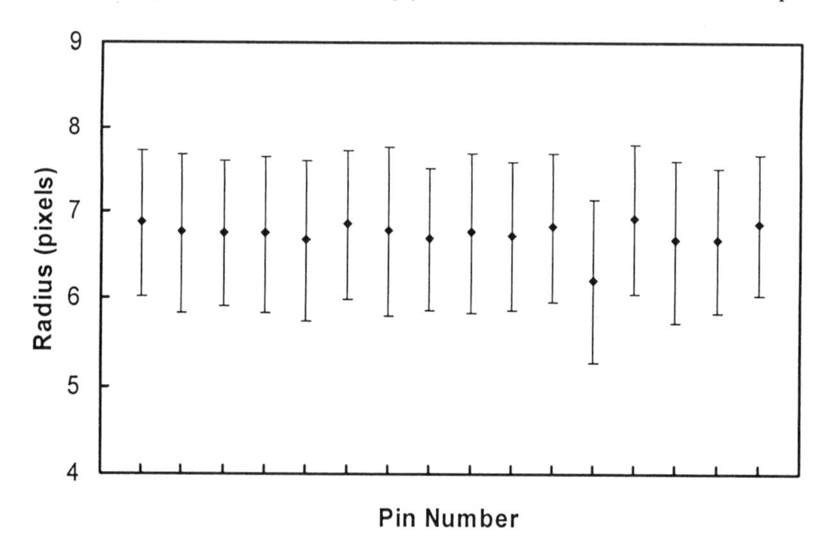

Figure 3: Measured spot radius versus pin number. The error bars show one standard deviation of the measured values. Spot size was similar for all pins, although pin 11 showed a tendency to have smaller spot sizes.

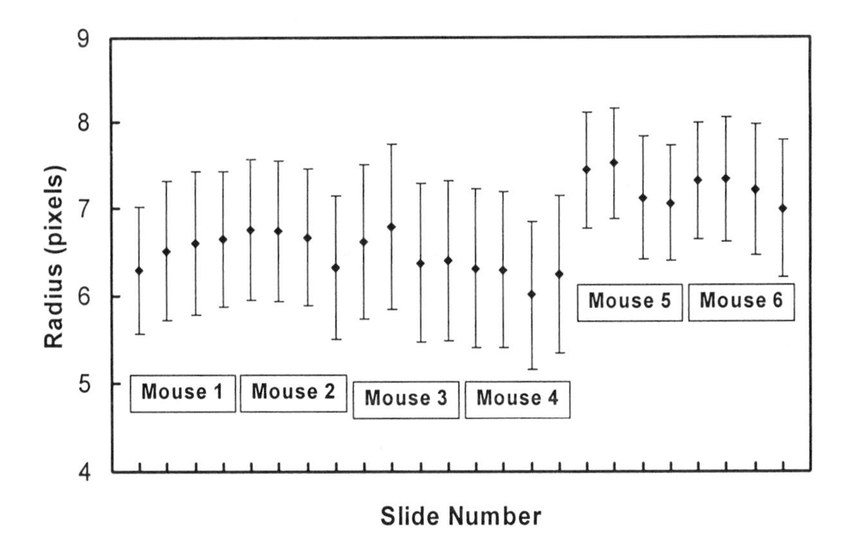

Figure 4: Measured spot radius versus slide number. The error bars show one standard deviation of the measured values. The slides associated with the last two mice had larger spot sizes.

and showed little dependence on slide or pin.

- Variation in separabilty was associated mostly with individual spot variation (83% in one channel, 90% in the other). Some differences between individual slides and pins were noted, but nothing extreme was found.
- Uniformity measures showed considerable variation by slide, largely due to the last eight slides, again, those from mice 5 and 6. These last eight slides had uniformity measures 40-50% worse than the rest of the slides in *both* channels (Figure 5). This supports the observation that something in the process changed after completion of the first four mice.
- The parameter for the intensity reduction in the low MRF region for spots was strongly dependent on the printing pin. This is another measure of uniformity, but is more sensitive to large low-intensity regions than the other pixelwise statistical measure. Further, we found that the variation between groups of slides was considerable (Figure 6), and essentially clustered the slides into three groups, divided by every two mice. The last two mice, again, suffered from greater spot nonuniformity than the others. This is readily apparent from the images, as we will show later.

- The sixteen MRF probability parameters were dependent on pin, supporting our decision to model the MRF parameters individually for each pin.

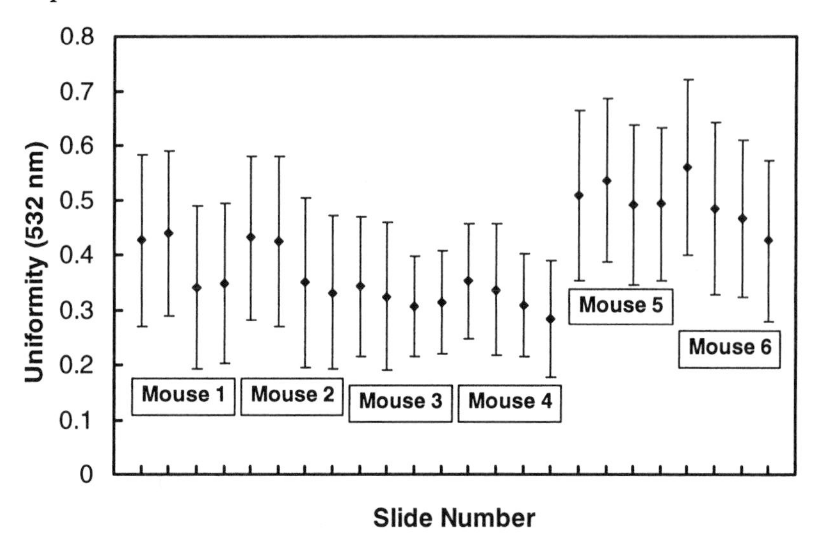

Figure 5: Uniformity of spots in the 532 nm channel image versus slide number. The error bars show one standard deviation. The spots on slides associated with mice five and six were somewhat less uniform than the others. Results for the 635 nm channel were similar.

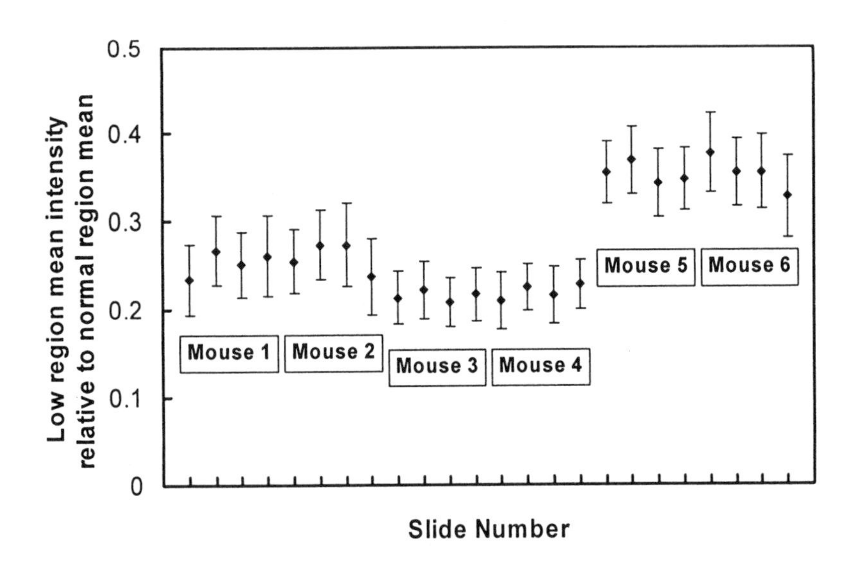

Figure 6: Spot defect region intensity versus slide number. The error bars show one standard deviation. Spots from slides associated with mice five and six have defects with greater intensity reduction, relative to the rest of the spot, than spots on the other slides.

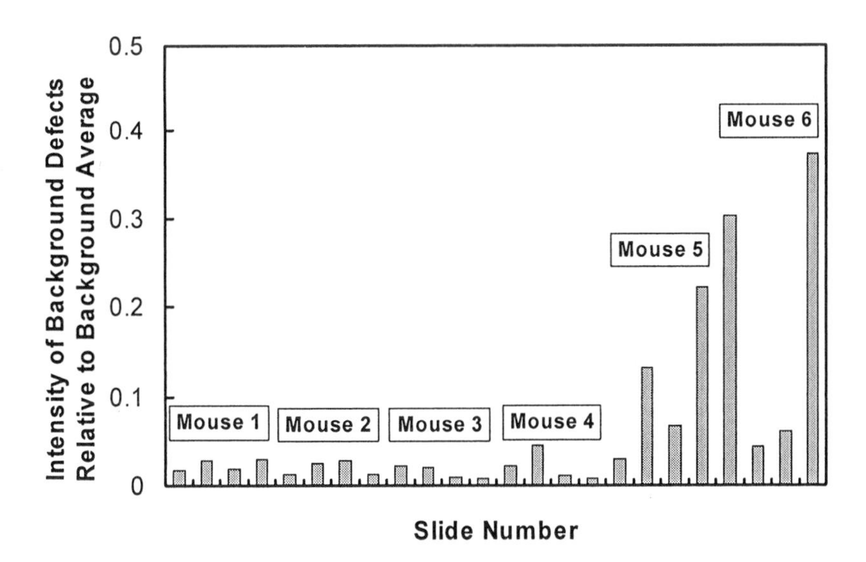

Figure 7: Background defect intensity versus slide number. Some of the slides from mice five and six had extensive background defects.

- The intensity of background defects as measured by the background MRF showed that several of the last eight slides had considerably more intense defects (Figure 7). Also, the probability parameters showed that the last eight slides were more likely to have background defects than the others. Observation of the images shows that the last eight slides generally had more extensive defects than the others. Once again, we interpret this as an indication that the last eight slides may have been handled differently than the rest.
- None of the measures showed strong slide-pin interactions in the ANOVA analysis. All accounted for less than 5% of variance, so this supports our decision to model slide and pin effects separately.

Results of the characterization strongly suggest that slides from mice five and six differ from the others in certain properties: their spots are larger and less uniform with more intense defects, their backgrounds are more likely to have defects, and such defects are more intense. This is apparent from observation of the slide images themselves, examples of which are shown in Figure 8. We note that the example from mouse 5 suffers from considerably more broken spots, and the breaks are more severe. Also, the slide from mouse five suffers from a large background defect in the lower right. The characterization measures above serve to detect and quantify these slide properties.

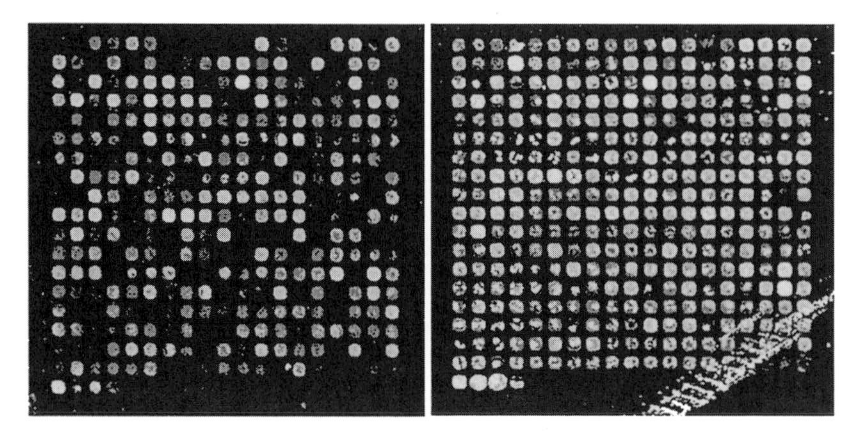

Figure 8: Excerpts from single-channel (532 nm) images for slide 2 (Mouse 1), left, and slide 19 (Mouse 5), right, illustrating differences in spot and background properties.

3.2 Simulation

As proof of concept, we present examples comparing simulated images in Figure 9. Two simulations are shown: one using parameters from the characterization of slides from the first four mice, and one using parameters from the characterization of slides from mice five and six. These correspond to the two cases shown in Figure 8. Note that the simulation based on parameters from mice five and six shows more broken spots and more

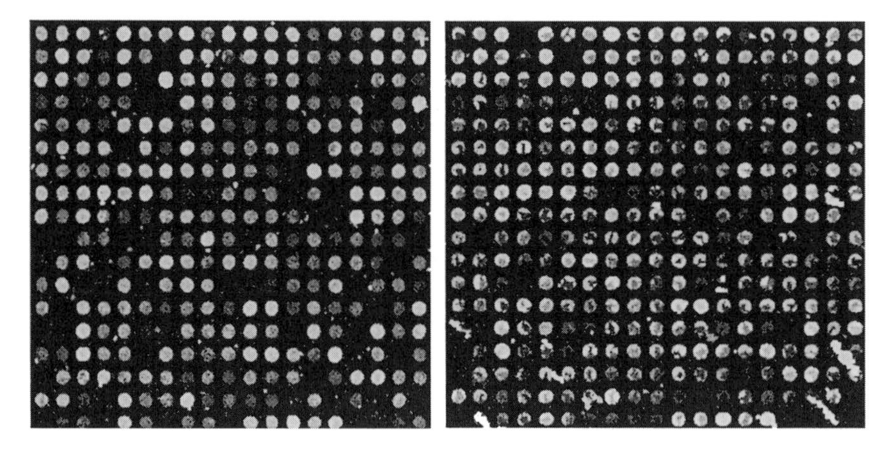

Figure 9: Excerpts from simulated images. The image on the left was produced from parameters derived from the slides for mice 1-4. The image on the right was produced from parameters derived from slides for mice five and six. The two are excerpted from the same region of the images; i.e., the same genes are shown. Note the increase in spot breaks and background defects in the second image.

background defects as expected from the differences in parameters that

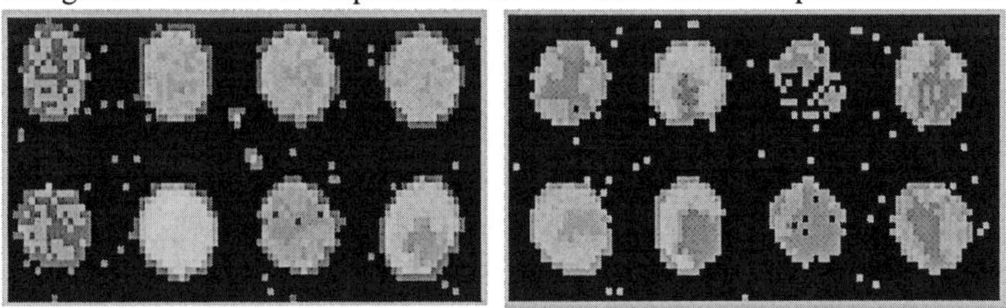

Figure 10: Close-up views of selected spots in the simulation images. The left side is taken from a simulation based on slide properties from mice 1-4; the right, from mice five and six.

control these properties. Close-ups of some of the simulated spots (Figure 10) show the variations in detail possible with the simulation techniques used.

4. DISCUSSION

Characterization of the images is not only useful for purposes of generating parameters for a simulation model as we have used here, but is also useful as a quality control procedure. Our characterization shows dramatically that the slides from the last two mice had important differences from the rest in terms of spot size, spot uniformity, and background defects. The impact on the biological observations in the original paper [Pritchard, *et al.*, 2001] is unclear. This could indicate that at least some of the mouse-to-mouse variation reported is not truly due to variations between mice but also due to variations in the treatment of the slides associated with those mice.

Quantifying the errors due to slide handling is a difficult matter. As yet, we do not have a good understanding of how things like spot uniformity and handling defects can affect the data, and, of course, different processing methods will have different sensitivities to these problems. Simulation can play an important role in evaluating these factors. With our simulation, we can hold most effects fixed, vary one, and see how the one parameter affects the final result. In the case of the mouse data, we can generate replicates of slides differing only in spot uniformity or only in background defect intensity, and get at least a rough estimate of the degree to which these problems might have affected the experimental results.

Like any simulation, ours is not perfect and it probably oversimplifies many effects. For example, our model does not model large scratches such as those seen in some of the images (Figure 8, right). Rather, background defects tend to be shorter, and distributed around the slide. Also, the simulated defects tend to travel in a southeasterly direction, a property of the causal MRF model we have used here. We expect our simulation to be revised in the future to improve our models for background effects and to add other effects. For example, multiscale modeling can help us to account for large defects, and the use of noncausal MRFs will remove the directionality of the defects.

We must note that the multivariate normal model of biological variation in gene expression is mostly a computational convenience and probably does not reflect the true behavior of gene-gene interactions. While this model should be adequate for many types of simulation experiments, users should be careful to consider the implications of its use on the experiments being performed. In the future, we expect to implement more complex and, hopefully, more accurate models.

Our image simulation is somewhat less detailed at the image level than other recent work intended for evaluation of image analysis algorithms [Balagurunthan, *et al.*, 2002]. On the other hand, our approach attempts to model the entire microarray system and is intended for generating large numbers of replicates for evaluation of microarray data analysis methods. While the purposes of the two approaches differ, it is likely that elements of each will be incorporated in realistic simulations of microarray image data in the future

We conclude that characterization of microarray images using a suitable model can reveal a great deal of information not found in the processed data. The results of characterization can be useful in quality control of major projects, and can identify potential sources of error. Further, the use of simulation could help to quantify the degree to which such problems may have affected final results.

5. SOFTWARE

The characterization program and the simulation program are written in C++ and are available directly from the author (lalush@statgen.ncsu.edu).

ACKNOWLEDGMENTS

This work was funded by a Career Development Award (K01 HG024028) from the National Human Genome Research Institute.

REFERENCES

Balagurunthan Y, Dougherty ER, Chen YD, Bittner ML and Trent JM (2002) Simulation of cDNA microarrays via a parameterized random signal model. J Biomed Optics 7: 507-523.

Besag J (1974) Spatial interaction and the statistical analysis of lattice systems. J Royal Stat Soc B 36: 192-226.

Cui X, Kerr MK and Churchill GA (2002) Data transformations for CDNA microarray data. Statistical Applications in Genetics and Molecular Biology (submitted).

Dobbin K and Simon R (2002) Comparison of microarray designs for class comparison and class discovery. Bioinformatics 18: 1438-1445.

Jain AK (1989) Fundamentals of Digital Image Processing (Englewood Cliffs, NJ: Prentice Hall).

Kerr MK and Churchill GA (2001) Statistical design and the analysis of gene expression microarray data. Genet Res 77: 2001.

Kerr MK, Martin M and Churchill GA (2001) Analysis of variance for gene expression microarray data. J Comp Biol 7: 819-837.

Lonnstedt I and Speed T (2002) Replicated microarray data. Stat Sinica 12: 31-46.

Pritchard CC, Hsu L, Delrow J and Nelson PS (2001) Defining normal variance in mouse gene expression. Proc Natl Acad Sci 98: 13266-13271.

Schena M, Shalon D, Davis RW and Brown PO (1995) Quantitative monitoring of gene expression patterns with a complementary-DNA microarray. Science 270: 467-470.

Szabo A, Boucher K, Carroll WL, Klebanov LB, Tsodikov AD and Yakovlev AY (2002) Variable selection and pattern recognition with gene expression data generated by the microarray technology. Math Biosci 176: 71-98.

Tseng GC, Oh M-K, Rohlin L, Liao JC and Wong WH (2001) Issues in cDNA microarray analysis: quality filtering, channel normalization, models of variations and assessment of gene effects. Nucl Acids Res 29: 2549-2557.

Wolfinger RD, Gibson G, Wolfinger ED, Bennett L, Hamadeh H, Bushel P, Afshari C and Paules RS (2001) Assessing gene significance from cDNA microarray expression data via mixed models. J Comp Biol 8: 625-637.

Yang YH, Dudoit S, Luu P, Lin DM, Peng V, Ngai J and Speed TP (2002) Normalization for cDNA microarray data: a robust composite method addressing single and multiple slide systematic variation. Nucl Acids Res 20: e15.

6

TOPOLOGICAL ADJUSTMENTS TO GENECHIP EXPRESSION VALUES

A search for hidden patterns in intensities

Andrey Ptitsyn
Pennington Biomedical Research Center, Baton Rouge, LA.

Abstract: In this paper we identify hidden patterns of intensities in Affymetrix biochips. The study shows that not only defective, but also "good" quality chips have local areas of elevated or lowered signal intensities. These areas may vary between chips, but have a distinct pattern, consistent for a particular chip type and batch. We describe an algorithm for quick estimation of Affymetrix biochip integrity either for a single chip or in a series of experiments. We also suggest a practical approach for improving the estimation of gene expression level by scaling probe intensities to the local grid mean intensity

Key words: Biochips, quality control, normalization, hidden patterns, shuffling

1. INTRODUCTION

Affymetrix oligonucleotide arrays are widely used to measure the abundance of mRNA molecules in biological samples [Lockhart *et al.*, 1996]. The old, traditional way to determine the gene expression from the measured intensities of individual probes is to calculate average distance between "perfect match" (PM) and "mismatch" (MM) probes

$$AverageDifference = \frac{\sum_{j \in A} (PM_j - MM_j)}{n_A};$$

Here A is a probe set with n_A number of probes. This method has been extensively criticized, particularly for the occasional "negative calls" [Chudin *et al.*, 2002; Zhou *et al.*, 2002] and in the latest versions of the

Affymetrix software, succeeded by the more stable Tukey's biweight method:

$$\log_2(Intensity) = Tukey_{biweight}(\log_2(PM_j - MM_j^*));$$

with MM_j* a version of MM_j that is never larger than the value of *PMj* [Affymetrix, 2002].

There are many other algorithms, based on PM only or including the mismatch intensity information [Chudin *et al.*, 2002; Zhou *et al.*, 2002; Li *et al.*, 2001]. These algorithms at some point, however, assume that the intensities of the probes across the chip are more or less uniform and can be used as a single probe set to determine the expression level [Hill *et al.*, 2001]. Affymetrix chips in the CAMDA '02 data set represent two variations of the technology. Human U95 chips have MM and PM probes placed in pairs with PM following MM directly below in the next row. PM-MM pairs of the same probe set are scattered throughout the chip. When their intensity values are gathered to estimate the expression of the gene, probes from the different corners of the chip are compared directly, presuming that the location within the chip does not affect the intensity. In the *E.coli* chips probes representing one gene are placed next to each other in the row. Corresponding "mismatch" probes are placed in the next row. All probes representing a single gene are always in proximity to each other. But uniformity of the chip is still assumed in the estimation of the background (is it valid to assume that background intensities are the same at each corner of a chip?) or in normalization of the gene expression values in one chip.

While considering probes scattered across the biochip, we should also take into account local effects. In case of defective chips such effects are obvious to the unaided eye. Large scratches, bubbles, "snow" or partially bleached chips are typically sorted out as defective. "Good" slides usually show no obvious defects and are retained for further analysis. But the question still remains. Are there any hidden patterns in what seems to be a randomly speckled field? How uniform is this signal intensity and how can this uniformity be measured and compared between the chips?

2. RESEARCH

In this study we presume that probes from each probe set are scattered randomly across all of the U95 (or similar) chip, as described by Affymetrix. Consequently, even if some genes are expected to be expressed at constantly high or constantly low levels, the intensity of the PM signals should be approximately equal in every part of the chip. MM signals should also be equal as they are supposed to reflect the background and cross-hybridization. Additionally, MM probes are also scattered randomly following their respective PM probes. By dividing the chip into two equal parts A and B through any direction in its two-dimensional plane, each half still contains 640X640/2=204,800 probes. If our hypothesis about randomly scattered probes is correct, the mean intensity values for both parts should be equal. We can test this by applying the *t*-test:

$$t_{AB} = \frac{\overline{X}_A - \overline{X}_B}{\sqrt{\sigma^2_{AB}\left(\frac{1}{n_A} - \frac{1}{n_B}\right)}};$$

where $\overline{X}$ are the means and n is the number of probes in each (A and B) half of the slide. With such large numbers the degree of freedom for the tabulated *t*-distribution is infinite. The same is still true for the 544X544 probe *E.coli* chip. The probes in each set of this chip are not randomized, but follow each other in a row. Still, the probe sets themselves are not organized in a particular order according to their expected expression.

Our calculations show that all chips in the CAMDA '02 test data set satisfy this condition. If divided in half in any direction, the mean intensities for each half will be equal as measured by the *t*-test for all chips. The mean intensity of each half is also equal to the mean intensity of the total chip with a 0.05 confidence level. Further analysis shows that the consequent division into 4 parts (dissecting by horizontal and vertical lines into 4 equal quarters) also results in sections with mean intensity equal to that of the entire chip. Potentially, this division into smaller and smaller sections can be continued and each fragment of the chip should have the same mean intensity equal to the mean intensity of the entire chip as long as the fragments are of adequate size. At some point fragments would be too small to appropriately represent the whole chip and the means are likely to differ due to stochastic reasons.

To determine how small such fragments could be so that their mean differs from the whole chip, we apply the following procedure. First, control sets of probes along the borders of the chip are removed from consideration, each filled with the median intensity value from the same chip. All probes are then thoroughly shuffled across the chip (1000 random relocations for each probe). The chip is divided sequentially into square areas by grid size g, which has values running from 1 to the grid size of the chip itself (for example, 640 for U95). For each grid g every square within the grid is compared to the mean intensity of the whole chip (or g=1). In order to make local grid and overall means independent, the overall mean is recalculated without the grid area it is being compared to. For this purpose it's temporarily filled with overall chip median values, comparable to the procedure used with the Affymetrix control probe sets. If a single area is different by *t*-test from the whole chip the process of decreasing grid size stops. For different chips in the CAMDA '02 data set, this minimal grid value varies from 8 to 24.

The same procedure can be applied to the chips without shuffling. If the probes are indeed randomly scattered across the chip and no other local effects exist, the minimal grid value should be the same as that determined with shuffling. The results of computation are, however, rather different. For most of the chips, the uniformity (i.e. means equal to entire chip mean by *t*-test) is lost on a much coarser grid. For the chips in the CAMDA '02 dataset this grid value varies from 6 to 12.

We repeated the same calculation in 4 different variants – all probes; PM probes only, MM probes only and PM-MM, where each mismatch value is subtracted from the corresponding perfect match probe. The results are surprisingly similar for all variants. This means that the effect cannot be attributed to either cross-match or specific hybridization only. After the grid is determined, for each fragment the *t* value is subtracted from the tabulated Student distribution value with 0.05 confidence level and infinite degrees of freedom. This data can be visualized as a heat map, where color indicates the degree of difference from the whole chip mean intensity, as shown in Figure 1.

On the other hand, variance in the same grid shows little significant difference from the overall variance of the chip and generally doesn't follow the same pattern as the mean. This test has been done on all chips with standard Fisher test:

$$F = \sigma_A^2 / \sigma_B^2$$

Calculated values for the Fisher test are compared to the table values for a confidence level of 0.05. As shown in Figure 2, most local areas of the chip have the same standard deviation as the entire chip. Only in a few segments of the chip does the local standard deviation differ from that of the chip overall. These are the same areas, where the difference between local and general means is highest.

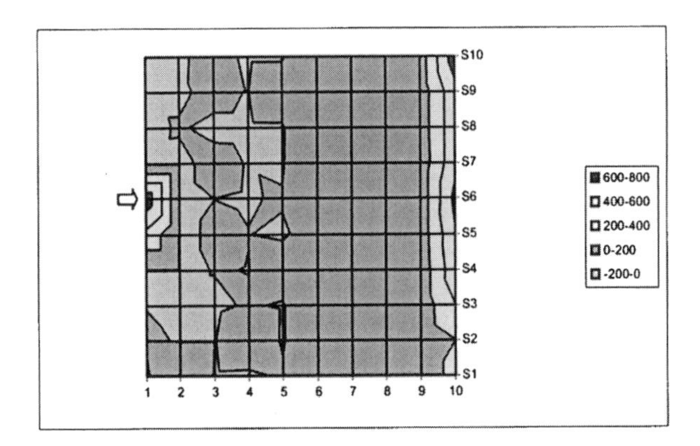

Figure 1. Heat map representation of intensity pattern from an Affymetrix E.coli chip, 000825_bgd_Ds1.CEL (CAMDA '02 data set). The grid is 10, which is the maximal grid size for which the shuffled probes' mean intensity in every square is equal to the overall mean intensity as measured by *t*-test. For raw, non-shuffled probes the grid is 4, i.e. mean intensity of at least one of 16 square sub-regions is different from the overall mean for this chip. The tone represents the difference in t value for each area of the grid from tabulated F distribution value with a 0.05 confidence level and degrees of freedom. In areas below 0 (see the legend) the means are equal, while other areas show significant deviation. The arrow marks the "belly button" present on all Affymetrix chips.

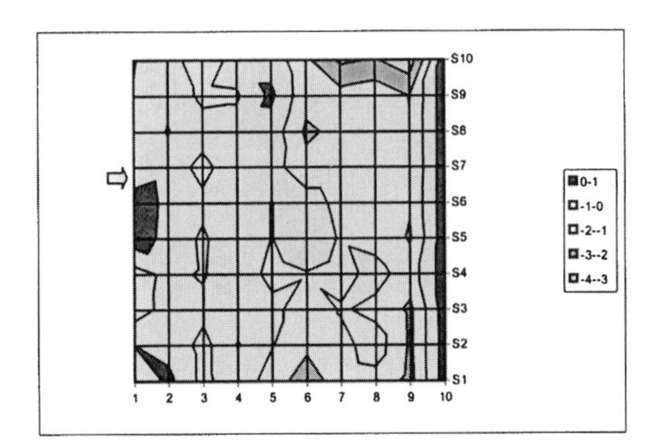

Figure 2. Heat map of significance of difference between local grid and overall standard deviations (Affymetrix E.coli chip, 000825_bgd_Ds1.CEL). The legend on the right shows the difference between calculated and table value for the Fisher test. With a 0.05 confidence level only a few areas differ from the overall standard deviation: a spot in the middle of the left side and the narrow strip along the right side. These areas are plotted with positive values.

3. DISCUSSION AND SUGGESTIONS

When presented as a heat map, the previously hidden pattern of intensity gradient becomes obvious. For most chips in the CAMDA '02 data set, this pattern suggests the diffusion of a dye in a fluid from a single source on one side of a chip towards the drain on the opposite side of the chip. The "holes" through which fluids are poured and drained are visible in the heat map. These are not necessarily related to injection/drain holes situated in the corners of the chips casing. From the information contained in the CEL files, we can only hypothesize about the internal design of the chip. The pattern visible on the Affymetrix chips can be explained by other reasons not involving the flow of fluids, for example, by specific placement of probes. Even if this pattern was created deliberately, the probe intensities should be adjusted for such local effects. A relatively small area of the chip has a mean intensity equal to that of the whole chip. All other areas show significant deviation in both PM and MM intensity. Now, knowing the value of deviation that is significant (assuming 0.05 confidence, used to plot the "significance map" in Figure 1) we can simply plot the average intensities for each part of the chip with a grid g.

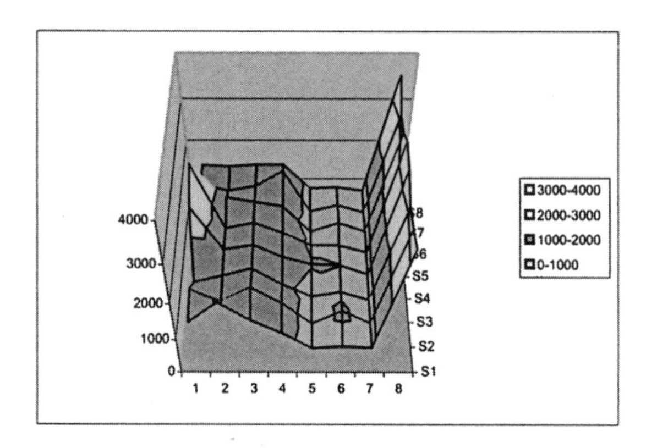

Figure 3. A different representation of the same 000825_bgd_Ds1.CEL chip (see Figure 1). This surface shows only the raw mean intensity for each square of the grid. Comparing it to the map on Figure 1 we can say how significant the deviation from the overall mean value is for each part of the chip.

The pattern shown in Figure 3 can possibly be explained by superposition of two gradients. The first gradient appears when the sample is injected. The intensity produced by this injection is highest near the site where the fluid enters the chip and gradually decreases towards the drain. The second gradient reflects the results of the subsequent washing. It reduces intensities around the injection site to a steep outcrop, but the washing effect also drops off towards the drain, leaving an elevated intensity along the far side of the chip. All chips in the CAMDA '02 data set show an intensity pattern consistent with this hypothesis. Needless to say, any deviations from the standard protocol are bound to be reflected in this pattern. A separate series of experiments would be required to determine how a change in temperature, concentration of reagents, or time would affect the gradients of intensities.

The experiments in the Affymetrix data set for CAMDA '02 were designed and conducted for a very specific purpose. From the "technology control" point of view, these experiments are essentially repetitions performed under standard conditions within a short period of time and using the same biological sample. Under these conditions only a negligible number of probe sets are expected to change intensity following the change in mRNA concentration in samples. We cannot pinpoint specific causes of variation in the data set. Nevertheless, some of the chips show significant differences from the others in the same series. For example, chip 001003_bgd_Ds5 (*E.coli*) has a significantly less uniform distribution of intensities. In this particular case the difference between the shuffled and non-shuffled grid is 6, while for other chips in this batch the difference is 4. The record in the table from the worksheet named "**Deviations**" within the excel file ***Algo_Dev_Ecol.xls*** reads "Left corner has large particle covering b3303_rpsE & b3340_c_fusA". This observation leads to the first practical application of this research. The difference between the maximal grid with and without shuffling can be used as a first rough estimation of the quality of a particular chip or to identify quickly the stray chips in the series of experiments.

The second practical application is more obvious. If different areas of a chip on each grid g are significantly different from the whole chip average, the probes from the same probe set should not be treated as if they are sampled from the same population. Before these probes are used to estimate expression of a particular gene, each probe's intensity must be adjusted to the local effect or gradient of average intensity. We suggest scaling each probe to the local mean intensity of the grid.

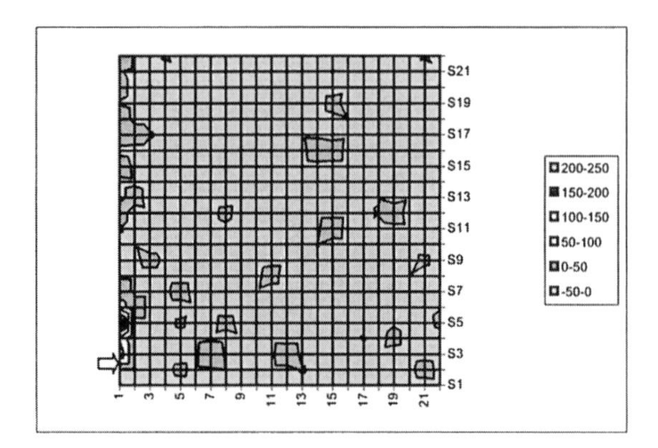

Figure 4. Heat map representation for grid *t*-tests for Affymetrix U95 chip 2353q99hpp_av08.CEL. As in Figure 1, gray areas (negative values) of the local grid mean are not significantly (conf 0.05) different from the overall mean for the chip. The U95 chips are much more uniform, but high local effect areas still account for 10 to 20% of the chip surface and have to be considered in estimation of gene expression level.

As it was already mentioned above, there is a significant difference in probe layout between the U95 and the 1286 (*E.coli*) chips. The U95 probe sets are scattered randomly across the chip, while in the 1286 they follow each other in a single row. As a result, the intensity is much more uniform in U95 chips. The second reason for the better consistency of U95 may be a result of differences in genome functioning and composition between prokaryotes and human. In human samples a much lower percentage of genes are likely to be highly expressed at a given time due to a larger genome size and specialization of tissues. High numbers of low-expressed and silent genes shift the overall mean expression towards the lower values and thus make the chip more uniform. Both shuffled and non-shuffled grid numbers for U95 are higher than those for 1286. Nevertheless, the 2-dimensional heat map plot shows a distinctive pattern (Figure 4). There are fewer grids, in which the local mean intensity is significantly (0.05) different from the overall mean intensity than in the 1286 (*E.coli*) experiments. However, such areas still account for 10% to 20% of the chip. The pattern is consistent through all U95 chips; the most affected area is the side near the injection port and a big "island" in the upper right quarter. The U95 chips have a relatively small number of grids that are significantly different from the overall mean. This number, or percent of the total surface affected by local differences, can also be used as a rough estimation of quality for the U95 and similar chips.

The approach discussed in this paper can, potentially, produce better results if applied to the images instead of the processed CEL files, but only

CEL files were available in the CAMDA data set. On the other hand, the fact that hidden patterns are revealed in the processed files demonstrates the utility of this approach for the purpose of quality control.

ACKNOWLEDGMENTS

The author would like to thank Dr. Julia Volaufova, Dr. Dawn Graunke and Dr. Simon Lin for their fruitful discussions of this work.

REFERENCES

Affymetrix: *Affymetrix GeneChip Expression Analysis Technical Manual. Santa Clara: Affymetrix;* 2002

Chudin E, Walker R, Kosaka A, Wu SX, Rabert D, Chang TK, Kreder DE. Assessment of the relationship between signal intensities and transcript concentration for Affymetrix GeneChip arrays. Genome Biol. 2002;3(1)

Hill AA, Brown EL, Whitley MZ, Tucker-Kellogg G, Hunter CP, Slonim DK. Evaluation of normalization procedures for oligonucleotide array data based on spiked cRNA controls.Genome Biol. 2001;2(12).

Li C, Wong WH., Model-based analysis of oligonucleotide arrays: expression index computation and outlier detection. Proc Natl Acad Sci U S A. 2001 Jan 2;98(1):31-6.

Lockhart DJ, Dong H, Byrne MC, Follettie MT, Gallo MV, Chee MS, Mittmann M, Wang C, Kobayashi M, Horton H, Brown EL: Expression monitoring by hybridization to high-density oligonucleotide arrays. *Nat Biotechnol* 1996, 14:1675-1680

Zhou Y, Abagyan R.Match-Only Integral Distribution (MOID) Algorithm for high-density oligonucleotide array analysis. BMC Bioinformatics. 2002;3(1):3.

SECTION IV

NORMALIZING RAW DATA

7

COMPARISON OF NORMALIZATION METHODS FOR cDNA MICROARRAYS

Liling Warren, Ben Liu
Bio-informatics Group Inc., Cary, NC 27511

Abstract: In a study done by Pritchard *et al.* [2001], normal gene expression variation was examined in six genetically identical male mice, to determine a baseline variation for gene expression studies in mice. In this paper, we use data from their study to accomplish the following three goals:

1) Evaluate five data normalization procedures along with two methods omitting data normalization, and study their impact on identifying baseline differentially expressed genes;

2) Perform pair-wise comparisons using McNemar's tests on five normalization methods and two methods omitting the normalization step;

3) Address data quality issues and examine the effect of normalization on analysis results for genes that do not meet either or both of the two data quality criteria.

Depending on which normalization method is used, whether omitting the normalization step or not, the number of genes and the set of the genes identified as differentially expressed from the same study can be substantially different. Analysis demonstrates that when data quality is not ensured, performing normalization can add noise to the data and can bias gene-based ANOVA results. Thus we conclude that ensuring data quality and establishing quality control measures is crucial to increase the effectiveness of normalization procedures and the accuracy of data analysis results. The study also reconfirmed that proper experimental design and establishing rigorous data quality control standards are indispensable factors for the success of a microarray experiment.

Key words: DNA microarrays, gene expression, data normalization, ANOVA, ANOCOVA, data quality control.

1. INTRODUCTION

Data normalization means removing certain systematic variation that is introduced during different steps in a microarray experiment, from microarray fabrication and biological sample preparation, to data acquisition. It is an intermediate step that has been routinely carried out after image analysis, and before data analysis, in a typical gene expression microarray experiment. Sources that can contribute to the signal intensity variation among array features include, but are not limited to:

1. The number of copies of DNA or RNA complementary to probes residing on a feature of a microarray;
2. Variation caused by substrates;
3. Efficiency of dye incorporation for different target strands of DNA or RNA in hybridization solution, and fluorescent signal detection during data acquisition;
4. Hybridization efficiency for different targets and in different hybridization solutions;
5. Variation caused by pins depositing probes onto an array.

Obviously, (1) is of biological interest. Its variation across treatments can be used to answer questions of biological interest. Variation from (2) - (5) has no biological significance and should be removed. Systematic factors can cause variation between features and between arrays. Normalization procedures try to remove such systematic variations. In this paper, we will compare results from adapting various normalization methods, and study how they impact identifying genes that are normally differentially expressed, based on the experiment done by Pritchard *et al.* [2001]. Particularly, we will perform pair-wise comparisons using McNemar's tests on five normalization methods and two methods omitting data normalization. Finally, we will explain the discrepancies caused by different normalization methods and make recommendations on which normalization methods to use for different study designs.

2. METHODS

The experiment by Pritchard *et al.* (2001) was done to study normal gene expression variation in the kidney, liver and testis tissues from six normal genetically identical male C57BL6 mice. Altogether, 24 cDNA based spotted microarrays were used for each of the three tissues. For each tissue, four replicate arrays were prepared for samples from each of the six mice. A common reference sample used in all arrays was made up of equal amounts of mRNA from each of the three tissues from each of the six mice. Within the four replicates for each mouse, two samples were labeled with green dye, Cy3 (G) while the reference samples were labeled with red dye, Cy5 (R), and the other two samples were labeled with Cy5 while the reference samples were labeled with Cy3. Each array was spotted with 5285 mouse cDNAs, 121 control and reference cDNAs. The control and reference cDNAs included yeast, human and E. coli genes. The normalization step was carried out to remove certain systematic variation that was introduced during a microarray experiment [Hedge *et al.*, 2000; Yang *et al.*, 2001]. During the array fabrication process, probes were deposited onto an array by a set of 16 pins. Pins were also referred to as blocks, corresponding to physical locations on an array. Based on the experimental design, genes were not replicated within blocks, or within arrays. Therefore, array effect, block effect, and any interaction terms involving these effects cannot be estimated in a gene-based ANOVA model. Data normalization was performed to remove these errors.

2.1 Method Description

Normalization method #1: Adjust for within block log ratio dependency on average intensity values using a local smoothing method, then adjust for between array dispersion by standardization.

This method is very similar to the one implemented by Pritchard *et al.* [2001]. To adjust the discrepancy in terms of dye incorporation between Cy3 (R) and Cy5 (G), a robust scatter-plot smoother lowess function is adapted to perform the local intensity dependent normalization [Dudoit *et al.*, 2000; Yang *et al.*, 2000]. Instead of using R and G, transformed coordinates M and A were used; where $M = \log_2 R/G$ and $A = 1/2(\log_2 R + \log_2 G)$ [Dudoit *et al.*, 2000].

We implemented this approach with the SAS procedure PROC LOESS (SAS Institute Inc., 1999b). In the LOESS procedure, a nonparametric method for estimating local regression surfaces is implemented. For example, let g be a regression function for y_i on x_i such that $y_i = g(x_i) + \varepsilon_i$, where ε_i is the random error term. The idea is that near $x = x_0$, a local approximation can be obtained by fitting the regression lines to the data points within a chosen neighborhood of x_0. The chosen neighborhood represents a specified percentage of all the data points around x_0. This can be controlled by the *smooth* option in PROC LOESS [Cohen, 1998]. It controls the "smoothness" of the estimated line. While implementing a normalization step with the local smoothing technique, we also wanted to see how different settings of this option would affect the normalization results, and ultimately influence the number of genes identified to be differentially expressed between mice.

When adjusting for log ratio dependency on average signal intensity within blocks, block means and array means are automatically adjusted to be at the same level, near zero. The next step is to adjust for data dispersion between arrays such that they are all on the same scale before applying gene-based ANOVA models. We have chosen to use array-specific standard deviation to perform such adjustment, similar to the one used in the paper by Pritchard *et al.*, where median of the deviation of the median was used.

Normalization method #2: Perform global normalization on log2-based ratios to adjust between block variation. Block specific mean is first substracted from each ratio value and then standardized by block-specific standard deviation. The number of blocks is 16 here, corresponding to 16 pins used in Stanford-type cDNA microarrays. After the normalization procedure, all log ratios are adjusted to mean zero and standard deviation 1. Since this step is performed at the block level, between array variation is thus automatically adjusted. Dependency of log ratio between two dyes on average signal intensity is not specifically adjusted, but dye effect is fit into the gene-based ANOVA model.

Normalization method #3: Perform global normalization on the logarithm of absolute value to adjust for between block variation. This method is very similar to method #2. The only difference is that the logarithm of absolute value instead of ratio is used for block-specific global normalization.

Normalization method #4: Use ANOVA to normalize the logarithm of absolute signal intensity values, then fit residuals from normalization ANOVA with a gene-based ANOVA to identify differentially expressed genes [Wolfinger *et al.*, 2001]. Based on the particular design in this experiment, the following mixed model for normalization was used:

$$y_{ijklm} = \mu + A_i + B_j + M_k + D_l + (AB)_{ij} + (AM)_{ik} + (AD)_{il} + \varepsilon_{ijklm} \qquad [1]$$

where y_{ijklm} is the log2-based fluorescence signal intensity measurement from array i, block j, mouse k, dye l and gene k. A is the main treatment effect for arrays, and B is the main effect for blocks or pins. This effect can be assessed globally since the same set of 16 blocks (pins) is used to deposit probes during array fabrication. M is the main effect for mouse, and D is the main effect for dye. Besides the main effects, we also fit some interaction terms that can contribute to the unwanted systematic variations. They are AB for array*block interaction, AM for array*mouse interaction and AD for array by dye interaction. It makes sense to fit the AD term here since dyes were swapped in four replicates for each mouse. If dye swapping were not done, this effect would be completely confounded with array*mouse, i.e., the AM effect. And finally, ε_{ijklm} is the random error term. Furthermore, *A*, *B*, *AB*, *AM*, and *AD* can be treated as the random effect. *M* and *D* can be considered as the fixed effects. To fit this mixed effect normalization model, the restricted maximum likelihood (REML) method was used. The Newton-Raphson method was used to estimate the variance components, which were then used to construct estimates for all random and fixed effects. The estimates for the fixed-effects were calculated using EBLUE (Empirical Best Linear Unbiased Estimation) and the random effects were computed using EBLUP (Empirical Best Linear Unbiased Prediction). We used the SAS procedure PROC MIXED to implement this normalization method [SAS Institute Inc., 1999b].

Normalization method #5: Use ANOVA to normalize the logarithm of the ratios, then fit residuals from normalization ANOVA with a gene-based ANOVA to identify differentially expressed genes. This method is almost identical to normalization method #4 except that log ratios are used instead of logarithm of absolute signal intensity values.

Method #6: Omit the normalization step, directly run gene-based ANOVA models as shown in equation [3], where y_{ijk} is the log ratio between Cy3 and Cy5 signal intensity measures.

Method #7: Omit the normalization step, directly run gene-based Analysis of Covariance models as shown below,

$$y_{ijk} = \mu + m_i + d_j + (md)_{ij} + \beta(x_{ijk} - \bar{x}_{...}) + \varepsilon_{ijk} \qquad [2]$$

where y_{ijk} is the logarithm of absolute signal intensity value for the *kth* replicate test sample in *ith* mouse with *jth* dye arrangement and x_{ijk} is the logarithm of absolute signal intensity value for the reference sample corresponding to y_{ijk}.

Methods #1 ~ #5 are normalization methods that have been commonly used for gene expression microarray experiments. Methods #6 and #7 omit the normalization step and their results were used to compare those from performing the normalization step. In other words, we wanted to see how the normalization procedure and how different normalization methods would affect the final gene-based ANOVA results. We also wanted to see what kind of difference it would make by using ratios between two dyes versus using the absolute values as the response variable. In addition, we wanted to examine how different local smoothing parameters used in the loess method would affect the analysis results.

2.2 Gene-based ANOVA models

After data normalization, the following gene-based ANOVA models were fit for each gene:

$$y_{ijk} = \mu + m_i + d_j + (md)_{ij} + \varepsilon_{ijk} \qquad [3]$$

where y_{ijk} denotes the log ratio for the ith mouse with jth dye arrangement in kth replicate, m_i and d_j represent the mouse and dye effects respectively, $(md)_{ij}$ is the interaction term between mouse and dye, and ε_{ijk} is the random error term. A significant mouse effect would indicate the gene is differentially expressed between mice. To correct for the multiple testing problem, we chose a similar level as the one used by Pritchard *et al.* [2001]. For the purpose of comparing various normalization methods, we think it is fair to use the same cut off value for all methods.

2.3 Evaluating normalization methods

The normalization step and gene-based ANOVA tests are two sequential steps during microarray data processing. Normalized results are used for further data analysis. So how data are normalized has a great impact on analysis results. The data used in this paper was generated using six genetically identical male mice and they were raised in a uniform environment. If there were genes differentially expressed, the differences must be due to factors other than DNA content. The purpose of the experiment was to define such baseline variation. Three very different criteria were used to choose genes as differentially expressed. The first one corresponds to alpha=0.05, knowing that 5 percent of genes are false positives. The second one uses conservative Bonferroni correction and the third one uses FDR (False Discovery Rate) method [Benjamini *et al.*, 1995; Efron *et al*, 2002]. Table 1 shows that there are genes identified as differentially expressed between genetically identical mice in all three tissues even when the most conservative Bonferroni correction is made.

Thus we have compared normalization methods based on the number of genes identified as differentially expressed genes.

Table 1. The numbers of genes detected by different normalization methods with three different criteria: (1) Raw_P: no adjustment for multiple testing, using 0.05; (2) Bonf_P: Bonferroni correction to ensure FWE (Family Wise Error rate = 0.05), using 0.05/number of genes; (3) FDR_P: using False Discovery Rate, method of Benjamini and Hochberg [1995] to adjust for multiple testing.

Tissue	Criteria	Method 1	Method 2	Method 3	Method 4	Method 5	Method 6	Method 7
	Raw_P	936	1440	1378	1544	1808	1253	1057
Kidney	Bonf_P	63	114	74	92	196	73	27
	FDR_P	488	1109	1044	1224	1551	757	441
	Raw_P	464	867	514	596	809	705	654
Liver	Bonf_P	12	31	13	10	25	0	1
	FDR_P	56	328	60	62	265	4	1
	Raw_P	853	966	733	596	825	3090	3042
Testis	Bonf_P	35	25	10	10	24	1956	1163
	FDR_P	272	407	153	62	232	3089	3038

We use McNemar's test [McNemar, 1947] to perform the comparison for a pair of methods. As shown in Table 2, a two by two table can be constructed for a pair of methods. Row marginal totals are the numbers of genes accepted by the first method, n_{A1}, and those rejected by the first method, n_{R1}. Similarly, column marginal totals are the numbes of genes accepted by the second method, n_{A2}, and those rejected by the second method, n_{R2}. The four cell numbers correspond to the number of genes accepted by both methods, n_{AA}; the number of genes accepted by the first method, but rejected by the second method n_{AR}; the number of genes rejected by the first method, but accepted by the second method, n_{RA}; and, the number of genes rejected by both methods, n_{RR}.

Table 2. McNemar's test comparison for a pair of methods

First Method	Second Method		
	Accept	Reject	Total
Accept	nAA	nAR	nA1
Reject	nRA	nRR	nR1

The McNemar test statistic can be constructed as:

$$\chi^2 = \frac{(n_{AR} - n_{RA})^2}{n_{AR} + n_{RA}}. \qquad [4]$$

Under the null hypothesis: the two methods declare the same number of genes as differentially expressed, the McNemar test statistic has a χ^2 distribution with 1 degree of freedom. A test statistic greater than 3.84 is considered to be significant at $\alpha = 0.05$. It means that the two methods being compared are statistically different in terms of identifying genes that are normally differentially expressed among the mice.

2.4 Assessing microarray data quality

To further examine why analysis results for some genes can be changed from being significant to non-significant or from non-significant to significant by either performing the normalization step or omitting it, data quality issues are addressed. Two data quality check criteria are established to identify genes that did not have good data quality.

2.4.1 Data quality check criteria #1

For each microarray experiment, the total mRNAs from the test sample and the reference sample were simultaneously hybridized onto the array. The common reference sample for all 72 microarrays was made up by an equal amount of mRNA from each of the three tissues of each of the 6 mice. On a gene-by-gene basis, it is expected that the sum of the amount of mRNA from all test samples should add up to the sum of the amount of mRNA in all reference samples. Let x_{ijk} be the amount of mRNA in tissue *i*, mouse *j* and replication *k* from the test sample and y_{ijk} being the amount of mRNA in tissue *i*, mouse *j* and replication *k* from the reference sample, then we would expect

$$r = \left(\sum_{i=1}^{3}\sum_{j=1}^{6}\sum_{k=1}^{4} x_{ijk}\right) \Big/ \left(\sum_{i=1}^{3}\sum_{j=1}^{6}\sum_{k=1}^{4} y_{ijk}\right) \qquad [5]$$

to be around 1. Figure 1 shows the histogram of r for all of the genes. When r is outside of the range between 0.5 and 2, we consider the hybridization data for that gene not reliable. The normalization effect was further examined for genes that did not meet this data quality check criterion. The thresholds 0.5 and 2 are somewhat arbitrarily chosen for this data set. It is not a stringent quality control criterion. Ideally, we would like to set up a much more stringent QC standard. More experimental data need to be evaluated to assess how a highly stringent criterion can be achieved.

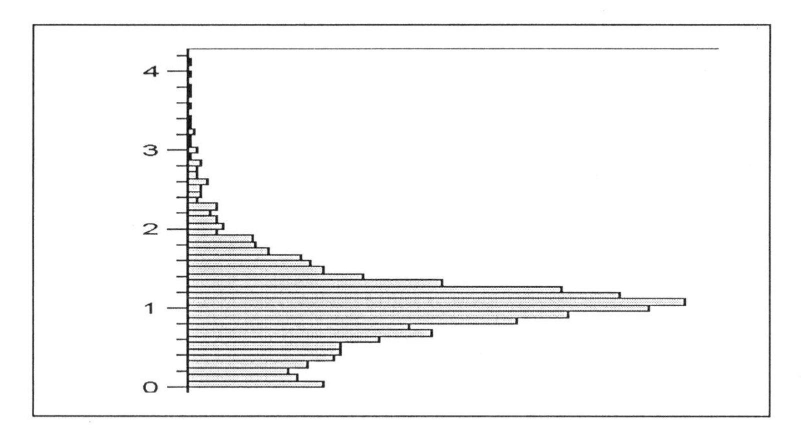

Figure 1. Histogram for r (the ratio between total mRNA in test samples and reference samples)

2.4.2 Data quality check criteria #2

For each gene on a microarray, the hybridization signal is calculated as the difference between the observed foreground signal and the background signal. This value can be anywhere from 0 to 65000. This number represents the amount of mRNA that is hybridized with the probes deposited on the microarray, with the assumption that the probes are in excess of the target molecules. When the hybridization signal value is very low, it is questionable whether the amount of mRNA is low in the test sample or the efficiency of the hybridization reaction is low. Without replicated data, it is difficult to distinguish between the two. In this experiment, four replicates were done for each gene. When we observe that one of the replicates had the signal intensity value much lower than the other three replicates, it is an indication that the hybridization reaction did not happen efficiently on that spot. If two very small numbers were observed from both the test sample and the reference sample for a gene, then it is reasonable to think that the probes were not deposited on the slide efficiently. If only one very small number was observed from either the test sample or the reference sample, it could be caused by inefficient labeling reactions for that gene. For a specific gene, when the value of signal intensity that was less than 100 observed among the four replicates, we consider the hybridization data for that gene not reliable.

The normalization effect was again examined for genes that did not meet this data quality check criterion.

3. RESULTS

3.1 Method comparison results

By applying methods #1 ~ #7 to the mouse kidney expression data, numbers of genes identified as differentially expressed by different methods are summarized in Table 3. From Table 3, we can see that depending on which normalization method was used, the results could differ significantly. Based on the number of genes identified as differentially expressed, the rank of the seven methods is: method #5, method #2, method #4, method #3, method #6, method #1, and method #7.

Table 3. The numbers of genes detected by method #1 ~ #7: Numbers on the diagonal are the numbers of genes detected by the specific corresponding method. Numbers off the diagonal in the upper triangle are those that are detected by either or both of the two corresponding methods, and the numbers off the diagonal in the lower triangle are the numbers of the genes that are commonly detected as differentially expressed genes by both methods.

	Method 1	Method 2	Method 3	Method 4	Method 5	Method 6	Method 7
Method 1	129	315	283	317	451	243	182
Method 2	89	275	414	451	522	362	318
Method 3	30	45	184	265	539	322	246
Method 4	38	50	145	226	561	344	277
Method 5	80	155	47	67	402	409	410
Method 6	51	78	27	47	158	165	174
Method 7	32	42	23	34	77	76	85

Normalization method #5, where log ratios were used, is similar to normalization method #4 where absolute values were used instead. We see that the log ratio based normalization method detected more genes than the one using the absolute fluorescent signal intensity value based method. This is not surprising because there were no spot replications done within the slide. Using ratios does a good job in terms of adjusting for between spot variation. The same trend was observed between normalization methods #2 and #3. Many more genes were detected in method #2, where ratios were used, than in method #3, where absolute values were used.

As opposed to results from normalization method #5, the other extreme is method #7, in which case no normalization was performed and gene-based ANOVA tests were done directly using absolute signal intensity values. It is obvious to see that this method detected the least number of genes as differentially expressed. Interestingly, when ratios were used, as in the case for method #6, significantly more genes were detected, where again, no normalization step was performed.

Table 4 shows the pair-wise method comparison results for the seven methods using the McNemar's test. Each value in a cell is a χ_1^2 value calculated from the McNemar's test. At $\alpha = 0.05$ level, we see there is only one pair of methods that was not significant. All other test results rejected the null hypothesis: the two methods declare the same number of genes as normally differentially expressed genes. The pair of methods that were not significantly different from each other was method #3 and method #6. Method #3 corresponded to the block-specific global normalization using the logarithm of absolute signal intensities, while method #6 omitted the normalization step and used log ratios to perform the gene-based ANOVA tests.

Table 4. McNemar's tests for pair wise method power comparisons: numbers shown are values from McNemar's tests. A χ^2 value > 3.84 indicates the two methods declared significantly different number of genes as differentially expressed.

	Method 1	Method 2	Method 3	Method 4	Method 5	Method 6
Method 2	94.3					
Method 3	12	22.4				
Method 4	33.7	5.99	14.7			
Method 5	197.2	44	96.6	62.7		
Method 6	6.8	42.6	1.2	12.5	223.8	
Method 7	12.91	130.8	43.95	81.82	301.77	65.31

When a local smoothing method was applied during normalization to adjust for the dependence of log ratios on average signal intensity values, we have looked at how different choices of smoothing parameters could influence the analysis results. As shown in Table 5, numbers of genes detected are listed for different settings of the smoothing parameter, ranging from 0.1 to 1.0 with the increment of 0.1. Numbers on the diagonal are the numbers of genes detected with the specified smoothing option. Numbers off the diagonal in the upper triangle are those that were detected by either or both of the two corresponding parameter specifications and the numbers off the diagonal in the lower triangle are the numbers of the genes that were commonly detected by both specifications. As described in normalization method #1, the idea here is that near $x = x_0$, a local approximation can be obtained by fitting the regression surface to the data points within a chosen neighborhood of x_0. The chosen neighborhood represents a specified percentage of the data points around x_0, which is controlled by the *smooth* option in PROC LOESS.

The overall trend from Table 5 shows that as the value of the smooth parameter increases, the number of genes detected increases too. From the lower triangle below the diagonal, we can see that numbers of jointly detected genes are relatively unchanged as the value of the smooth parameter increases. This shows additional genes can be detected to be significant as more genes are included for local smoothing. The results shown here demonstrate that the local smoothing parameter plays an important role in the normalization step, thus directly influencing the results from the gene-based ANOVA tests. Although the use of a number between 0.1 and 0.3 is recommended [Dudoit *et al.*, 2000] for the smooth parameter, it is still a large range where very different numbers of genes can be detected with different settings of the parameter.

Table 5. The numbers of genes detected by different smoothing parameters: numbers on the diagonal are the numbers of genes detected with the specified smoothing option. Numbers off the diagonal in the upper triangle are those that are detected by either or both of the two corresponding parameter specifications and the numbers off the diagonal in the lower triangle are the numbers of the genes that are commonly detected by both specifications

	10%	20%	30%	40%	50%	60%	70%	80%	90%	100%
10%	62	110	143	166	176	182	190	186	193	203
20%	53	101	139	164	176	183	187	185	193	205
30%	48	91	129	158	173	183	187	188	197	210
40%	49	90	124	153	169	181	187	188	196	211
50%	48	96	118	146	162	174	184	186	194	212
60%	47	85	113	139	155	167	181	183	182	213
70%	46	88	116	140	152	160	174	177	187	209
80%	46	86	111	136	146	154	167	170	180	204
90%	45	84	108	133	144	151	163	166	176	204
100%	45	82	105	128	136	140	151	152	158	186

3.2 Data quality check results

In the kidney data, there are 388 genes that were identified as normally differentially expressed among the mice. Among these 388 genes, there are 156 genes that failed to meet data quality check criteria #1 and 124 genes that did not meet criteria #2. The normalization effect was examined for genes that did not meet either of the data quality check criteria. As shown in Figure 2, for genes that did not meet data quality check #1, P values from gene-based ANOVA tests before and after normalization method #1 were plotted. We can see that for these genes, normalization can change the gene-based ANOVA results either from being significant to non-significant or vice versa. The same trend was also observed in Figure 3, where a similar plot was done for genes that did not meet data quality check #2.

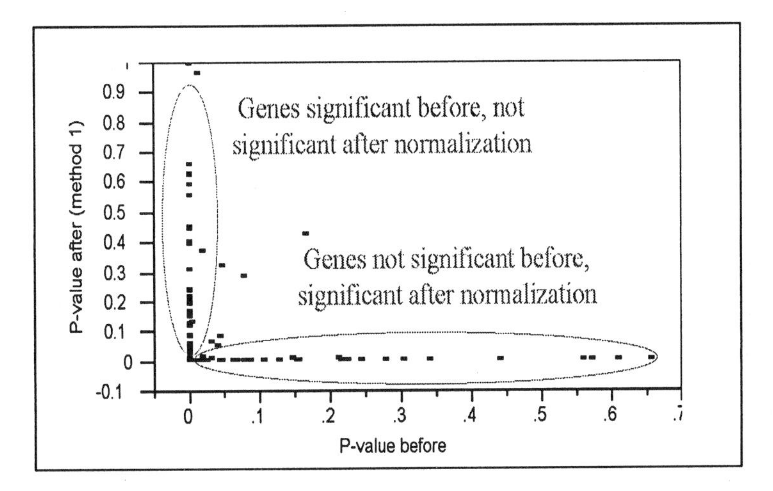

Figure 2. P-values for genes that do not meet data quality criteria # 1 before and after using normalization method #1.

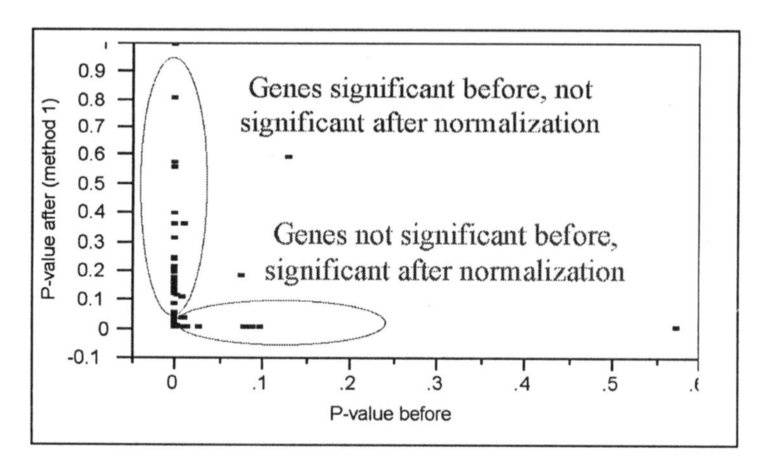

Figure 3. P-values for genes that do not meet data quality criteria # 2 before and after using normalization method #1.

The above results show that when the hybridization data are not reliable, the analysis results are questionable when data normalization is performed. With unreliable data, normalization can actually add noise to the data and bias the analysis results. For example, during the local smoothing step, genes that are not of good quality are locally grouped together with genes that are of good quality. During local smoothing regression model fitting, genes of bad quality can affect how the local regression curve is fit, and change the predicted value for all of the genes involved.

4. DISCUSSION

Systematic variation in gene expression microarray experiments caused by effects such as array and pin are obvious, but these effects can not be fit into the gene-based ANOVA tests to identify differentially expressed genes between mice due to the design of the experiment. Thus, implementing the normalization procedure is a necessary step. Another obvious effect that needs to be adjusted is the discrepancy between the incorporation of two dyes. A local smoothing technique has been adapted to perform such adjustment, while the choice of smoothing parameter has a great impact on variance structure after normalization. Among all the methods compared here, the ratio based ANOVA normalization procedure declared the most number of differentially expressed genes in mice. Since it is not known how many of the detected genes are true positives, and how many are false positives, definite conclusions can not be drawn from data analyzed here on which method is the best one.

By applying different normalization methods, different sets of genes can be identified as differentially expressed. To deal with this, it is important to keep in mind that a large-scale microarray experiment is just the starting point for further detailed experiments to answer specific questions of biological interest. Once we have identified a set of differentially expressed genes and there are discrepancies between using different normalization methods, one can include all the genes that have been detected regardless of the methods used for further experiments. If resources were limited, it would then make sense to include the conservative set of genes, that is, use the genes that have been jointly discovered by all different methods as new target genes for future studies.

From examining the effect of normalization for genes that are not of good quality, we can see that it is crucial to ensure data quality first, before normalization. When bad data are included, although normalization can remove systematic errors, it can also add noise to the data and change the analysis results significantly. Rigorous data quality control standards need to be established to screen the large amount of data produced from a microarray experiment. Using positive and negative controls can check hybridization data quality. Multiple labeling reactions can be also incorporated into the experimental design to account for the variation caused by dye labeling reaction variation.

For gene expression microarray experiments, there are many sources of variation that need to be controlled such that questions of biological interest can be addressed with formal statistical testing. With experimental designs, replication provides a way to estimate experimental errors, which is required for tests of significance and for confidence interval estimation. When a single replicate experiment is done, there is no way to determine whether observed differences indicate real differences or are due to inherent variation. Well recognized microarray experimental designs include reference design, loop design, and BIB (Balanced Incomplete Block) designs (Kerr *et al.*, 2001; Kerr *et al.*, 2001; Simon *et al.*, 2002). It is always a good idea to perform power analysis when choosing a proper design and deciding on factors such as number of samples, number of replicates, etc.

One extreme case we have seen in this data set is when method #6 was applied, in which there were an alarmingly large number of genes detected even if the extremely conservative Bonferroni correction is made. In method #6, no normalization was done and ratios were used to identify differentially expressed genes in testis tissue. After carefully examining original testis data, we noticed that mouse 3 and mouse 4 have very different ratio means at the array level from other mice. This either shows that the normalization step is crucial to adjust for the array effect or there might be data recording errors in the testis tissue expression data. It was later shown that there were data handling errors involved in the testis data set.

Molecular biology has come a long way from using one or few replicates to study one gene at a time. Microarray technology has brought much promise and excitement for genome-wide gene expression studies [Schena *et al.*, 1995; Shalon *et al.*, 1996]. The proper utilization of such technology, and close collaborations between biologists and statisticians, has been proven to be a key point towards ensuring the success of an experiment.

ACKNOWLEDGMENTS

We sincerely thank Dr. Bruce Weir, Dr. Ross Whetten and Dr. Yinghsuan Sun for their comments and suggestions.

REFERENCES

Benjamini Y, Hochberg Y (1995) Controlling the false discovery rate: a practical and powerful approach to multiple testing. Journal of the Royal Statistical Society 57:289-300.

Cohen RA (1998) An introduction to PROC LOESS for local regression. Paper 273. SAS Institute Inc., Cary, NC.

Dudoit S, Yang YH, Callow MJ, Speed TP (2000) Statistical methods for identifying differentially expressed genes in replicated cDNA microarray experiments. Technical report #578, Department of Statistics, University of California at Berkeley.

Efron B, Tibshirani R (2002) Empirical bayes methods and false discovery rates for microarrays. Genet Epidemiol 23(1):70-86.

Hegde P, Qi R, Abernathy K, Gay C, Dharap S, Gaspard R, Hughes JE, Snesrud E, Lee N, Quackenbush J (2000) A concise guide to cDNA microarray analysis. Biotechniques 29(3):548-50, 552-4, 556

Kerr MK and Churchill GA (2001) Experimental design for gene expression microarrays, Biostatistics, 2, 183-201.

Kerr MK and Churchill GA (2001) Statistical design and the analysis of gene expression microarrays, Genetical Research, 77, 123-128.

McNemar, Q (1947) Note on the sampling error of the difference between correlated proportions or percentages. Psychometrika 12: 153-157.

Pritchard CC, Hsu L, Delrow J, Nelson PS (2001) Project normal: defining normal variance in mouse gene expression. Proc Natl Acad Sci 98(23):13266-71

SAS Institute Inc. (1999b) SAS/STAT Software Version 8. SAS Institute, Inc., Cary, NC.

Shalon,D., Smith,S.J., Brown,P.O. (1996) A DNA microarray system for analyzing complex DNA samples using two-color fluorescent probe hybridization. Genome Res. 1996 Jul;6(7):639-45.

Schena M, Shalon D, Davis RW, Brown PO. Quantitative monitoring of gene expression patterns with a complementary DNA microarray. Science. 1995 Oct 20;270(5235):467-70.

Simon R, Radmacher MD, Dobbin K. (2002) Design of studies using DNA microarrays, *Genet Epidemio*, 23(1), 21-36.

Westfall, P.H. and Young, S.S. (1993) Resampling-based multiple testing: examples and methods for p-value adjustment. Wiley series in probability and mathematical statistics. Wiley. New York

Wolfinger,R.D., Gibson,G., Wolfinger,E.D., Bennett,L., Hamadeh,H., Bushel,P., Afshari,C., Paules,R.S. (2001) Assessing gene significance from cDNA microarray expression data via mixed models. *J Comput Biol.*, 8(6), 625-637.

Yang, Y.H., Dudoit, S., Luu, P., Speed, T.P. (2001) Normalization for cDNA Microarray Data, SPIE BiOS 2001, San Jose, California.

Yang,Y.H., Dudoit,S., Luu,P., Lin,D.M., Peng,V., Ngai,J., Speed,T.P. (2002) Normalization for cDNA microarray data: a robust composite method addressing single and multiple slide systematic variation. *Nucleic Acids Res.*, 30(4), e15.

SECTION V

CHARACTERIZING TECHNICAL AND BIOLOGICAL VARIANCE

8

SIMULTANEOUS ASSESSMENT OF TRANSCRIPTOMIC VARIABILITY AND TISSUE EFFECTS IN THE NORMAL MOUSE

Shibing Deng, Tzu-Ming Chu, and Russ Wolfinger
SAS Institute, Cary, NC

Abstract: We consider two linear mixed models for the normal mouse data [Pritchard *et al.*, 2001]. One models the $\log_2$ intensity measurements directly and the other models the $\log_2$ ratios. In each approach, we treat a mouse as a fixed effect, and alternatively, we also model it as a random effect to assess its variability directly. We compare the results from these mixed model approaches. The models agree that array variance is much larger than other sources of variability, but differ somewhat in their lists of genes exhibiting the most significant mouse effects. Under a Bonferroni criterion, the ratio-based model we consider produces more genes with significant mouse effects than the intensity-based model, but fewer genes with significant tissue effects. Both models demonstrate a general statistical framework for concurrently estimating sources of variability and assessing their significance.

Key words: Microarray, Mixed Model, ANOVA

1. INTRODUCTION

Microarray technology provides a great opportunity for functional genomics research. However, considerable variability in microarray data makes assessment of experimental effects challenging. Common sources of variation in the data include array, subject, treatment, dye, print-tip and gene. Pritchard *et al.* [2001] study variation of normal mouse expression in three different tissues using cDNA microarrays. They find a small fraction of genes (0.8-3.3%) exhibiting significant mouse variance in the three tissues. Their data set provides researchers an excellent opportunity to understand variability in a typical microarray experiment.

The experimental design of Pritchard *et al.* [2001] is a popular one, in which half of the experimental intensities arise from a common reference sample. While this design has been shown to be inefficient compared with more classical incomplete block or split-plot designs [Kerr and Churchill 2001], we are analyzing the data after the fact and so will try to model it appropriately.

A typical analysis proceeds by forming ratios within each spot and then adding these ratios to various methods. The ANOVA-based results from Pritchard *et al.* [2001] follow this line, and we greatly welcome more widespread use of classical ANOVA as well as more emphasis on variability assessment. In their ANOVA model, mouse variation is assessed by an F-statistic associated with mouse that is treated as a fixed effect, and analysis is conducted separately for the three tissues. In this paper, we approach the normal mouse data using a linear mixed model, a generalization of classical ANOVA with the following advantages:

- capability to directly model the $\log_2$ intensity measurements and to accommodate all known sources of variability across all of the microarrays;
- flexibility to specify both fixed and random effects, the latter enabling inferences on the former to be extended to general populations of interest;
- a comprehensive analysis framework for complex experimental designs and unbalanced data;
- extensive output statistics suitable for dynamic visual display, including those useful for quality control, inference, and classification.

The goal of this paper is to compare the models of log intensities with those of log ratios, and under both situations, to assess the magnitude and significance of known sources of variability. When modeling the $\log_2$ intensity measurements directly, there is the chance to gain some efficiency. The intuition here is that one sacrifices some valuable information in forming ratios. In particular, the fact that the same reference sample is used on each array can be exploited ipso facto to obtain a highly precise estimate of the reference sample expression before any relative comparisons are made. We accomplish this in our modeling context by assigning a unique treatment identifier to the reference sample, in addition to the three other identifiers for kidney, liver, and testis, and use a random effect to model spot-to-spot variability. Section 2 describes these two models and Section 3 contains results. Discussion is in Section 4 and conclusions in Section 5.

2. METHODS

For the log intensities, we follow Wolfinger *et al.* [2001] and Chu *et al.* [2002] and employ the following two statistical models in turn:

$$Y_{ijkdg} = M_i + T_j + D_d + TD_{jd} + MA_{ik} + MAD_{ikd} + e_{ijkdg} \quad (1)$$

$$R_{ijkdg} = M_{ig} + T_{jg} + D_{dg} + MA_{ikg} + MT_{ijg} + \varepsilon_{ijkdg} \quad (2)$$

Y represents the observed $\log_2$ intensity measurements, and *M, T, A, D* represent unobserved effects for mouse (i=1,...,6), treatment (j=kidney, liver, testis, reference), array (k=1,...,12, nested within mouse) and dye (d=Cy3,Cy5), respectively. Concatenation of these symbols represents interactions of the constituent effects. R in Model 2 denotes residuals calculated as observed Y minus fitted Y from Model 1, and the subscript g indicates that Model 2 is fitted separately for each gene. Both the normalization model (Model 1) and gene specific model (Model 2) assume a residual error from a normal distribution with zero mean and constant variance.

In Model 1 we fit all the data simultaneously. It evaluates the systematic across-gene effects in the data and provides a basic way to adjust for them. Here we assume mouse and array are random effects with normal distributions as follows:

$$M_i \sim N(0,\sigma_m^2), \qquad MA_{ik} \sim N(0,\sigma_a^2), \qquad MAD_{ikd} \sim N(0,\sigma_d^2).$$

Random effects provide more general inference possibilities than classical fixed-effects ANOVA models. Random effects allow one to make inferences to the entire population from which they are assumed to arise. They also enable direct estimation of the variance components associated with each random term.

The residuals from Model 1 are corrected for systematic across-gene effects and will be our normalized log intensity data for further analysis in Model 2, in which we fit a mixed effect model for each gene separately. We will first treat mouse as a fixed effect in Model 2 and array as a random effect (Model 2a) with $MA_{ijg} \sim N(0,\sigma_{ag}^2)$.

When mouse is a fixed effect, the variances of the mouse effect for each tissue are assessed from the *F*-statistics associated with the mouse by tissue interaction (MT) at each level of the tissue effect. It is essentially the test of mouse effect for each tissue. The *F*-statistic itself is a ratio of variation of mouse to variation of error. Thus, it can be used to test the significance of mouse-to-mouse variability.

Alternatively, we fit Model 2 with terms associated with mouse and array as random effects (Model 2b); that is, each is assumed to be normally distributed and independent of other random effects in the model:

$$M_{ig} \sim N(0,\sigma^2_{mg}), \qquad MA_{ikg} \sim N(0,\sigma^2_{ag}), \qquad MT_{ijg} \sim N(0,\Sigma_g).$$

Here Σ_g is a 4x4 diagonal matrix with σ^2_{Kg}, σ^2_{Lg}, σ^2_{Tg}, and σ^2_{Rg} along its diagonal. Each random effect has a zero mean and its own unique variance component, except for *MT*, which we assume has distinct variance components for each level of *T*. This enhancement to the standard model is included to facilitate comparisons with the results of Pritchard *et al.* [2001]. In this model, we can directly estimate the variance components and compare their magnitudes.

In both models (mouse as fixed effect and random effect), we can also assess the tissue effect. Although in this experiment it is not the main effect of interest, we use it as an example to demonstrate simultaneous evaluation of treatment effects and variability.

Generally, more data means more power for estimating parameters and testing hypotheses. By incorporating the reference sample into the model as an additional treatment level, we are able to estimate the variance components more accurately, since it has information about mouse, array, and replicate variability, even if it does not provide any information about the tissue effect directly.

Following Pritchard *et al.* [2001], we also consider the ratio of the two channels as a measure of relative expression levels of genes between treatment and control samples. In this approach, we have half the number of observations, and the gene-specific model does not have an array effect since there is only one observation per gene for each array.

$$Y_{ijkdg} = M_i + T_j + D_d + TD_{jd} + MA_{ik} + MAD_{ikd} + e_{ijkdg} \qquad (3)$$

$$R_{ijdg} = M_{ig} + T_{jg} + D_{dg} + MT_{ijg} + \varepsilon_{ijdg} \qquad (4)$$

Here, Y is the $\log_2$ transformed ratio of the two channels. All the other terms are the same as in Models 1 and 2, except *T* has only three levels here.

As a parallel study to our first approach, we will model the log ratio in two ways: treating mouse as a fixed effect (Model 4a) and as a random effect (Model 4b). To make the model more comparable with the Pritchard model, we further specify a heterogeneous residual structure, that is, each tissue is allowed to have a distinct residual variance parameter. This specification also provides a better fit than using a homogeneous variance structure based on model diagnostic criteria such as AIC (results not shown).

Restricted maximum likelihood (REML) is used to obtain the variance components estimates for the different models, and these are then used to construct generalized least squares estimates of all fixed effects in the model as well as their standard errors. All analyses are conducted using the SAS Microarray Solution software.

3. RESULTS

In the first approach, we fit a mixed model to the log intensities, considering mouse as a fixed effect (Model 2a). We estimate the variance of arrays and the residual errors for each gene. Figure 1 displays the distributional characteristics of these estimates. Note the array variance is distributionally an order of magnitude larger than the residual variance.

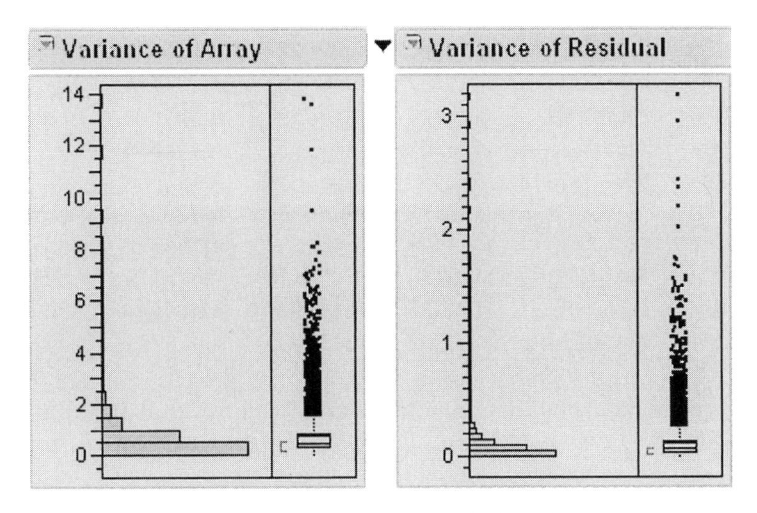

Figure 1. Variance components of Model 2a when mouse is fixed effect

To assess the mouse effect for each tissue, we create separate *F*-statistics from the corresponding rows of the *MT* effect contrast matrix (as implemented in the SLICE= option in the LSMEANS statement of SAS Proc Mixed). With a Bonferroni adjustment, we find 93 significant genes for kidney, compared to 127 found in Pritchard *et al.* [2001], with an overlap of 24 genes. We also find 20 and 11 significant genes for liver and testis, respectively, compared to 17 and 77 in the original paper, and the overlap is 5 and 0 genes, respectively. Table 1 lists the top 20 genes for each tissue. The reason for the small overlaps could be due to the modeling difference (intensity versus ratio) or to differences in data normalization.

Table 1. Top 20 genes with significant mouse effect for Kidney, Liver and Testis (Model 2a and Model 4a). Bold genes are found in Pritchard *et al.* [2001].

Model 2a			Model 4a		
Kidney	Liver	Testis	Kidney	Liver	Testis
AI452034	**AI385595**	AI428890	**AI428899**	AI427514	AI451732
AI428930	AI427514	AI449492	**AI449600**	**AI893426**	**AI447922**
AI413988	**NM_011817**	AI449974	**AI415173**	**AI385595**	AI449619
AI449691	AI451998	AI448808	AI448283	**AI426455**	**AI447316**
AI452254	AI528715	AI449619	**AI666732**	AI426672	**AI428919**
AI415173	AI413228	AI452034	AI427914	**AI528676**	AI414844
AI428457	AI448727	NM_010518	AI429463	**AI465375**	AI413204
AI450962	AI449336	AI447512	AI452132	**AI385595**	**AI428900**
AI427870	AI327095	AI464395	AI464346	**AI326008**	AI573427
AI451251	AI325092	AI324194	AI452338	**AI450826**	AI448386
AI452155	**AI893426**	AI428866	**AI325224**	**NM_013642**	AI448808
AI666521	AI465383	AI447752	AI465266	AI427478	AI661341
AI448388	AI893759	AI428784	AI464337	**AI324640**	**AI449611**
AI427914	AI894225	AI414363	AI449873	**AI450559**	AI414773
AI448964	**NM_013642**	AI893440	AI451616	AI452258	AI450852
AI451919	AI452034	AI465177	AI464459	**AI429284**	AI427146
AI448808	AI447880	AI449198	AI450559	AI464359	AI428907
AI448001	AI414295	AI449474	AI464413	AI528715	AI450318
NM_009406	AI451137	AI452035	**AI428930**	**AI450344**	AI429541
AI464337	**AI450826**	AI464567	AI428038	AI326797	AI449376

Over a thousand genes are identified with significant differences in expression between tissue pairs. Using these significant genes, the four tissues can be completely separated in clustering the standardized least squares means of the six mice (results not shown).

Next, we consider mouse to be a random effect in the same model as above to directly assess the mouse variance. Figure 2 displays the variance components from Model 2b. The array variance component *MA* has the largest values, followed by the tissue-specific *MT* components. Note the *MT* reference-sample distribution is approximately 5 times smaller than each of the tissue-specific components. This makes sense given that the reference sample is an equal mixture of all 18 mouse-tissue mRNAs. The residual error variance and the overall mouse variance component are moderate in size.

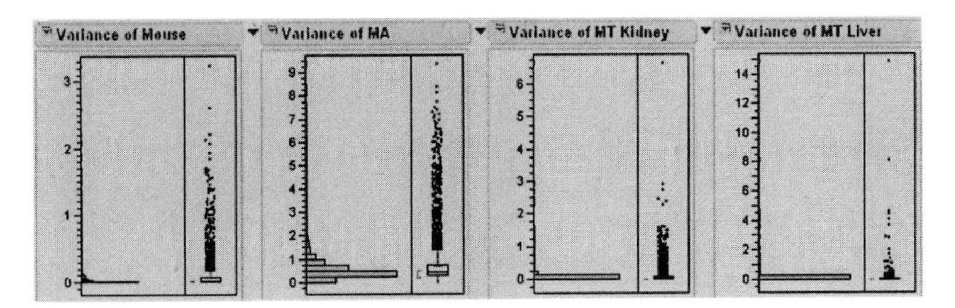

Figure 2a. Variance Components of Model 2b when mouse is a random effect

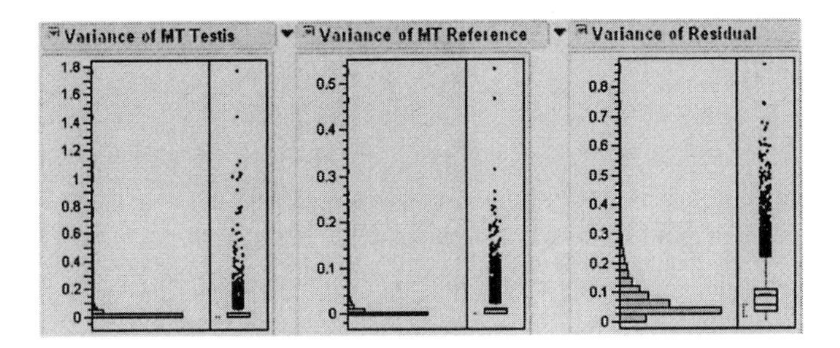

Figure 2b. Variance Components of Model 2b when mouse is a random effect

Figure 3 plots statistical significance (-$\log_{10}$ p-values) versus effects size ($\log_2$ fold change) for the three pairwise differences between tissues. Over 2000 genes have p-values passing the strict Bonferroni multiple comparisons criterion for at least one of the comparisons between the three tissue types. Note that direct differences with the reference sample are not necessary but are implicitly included because, for example, log(K/R) – log(L/R) = log(K) – log(L).

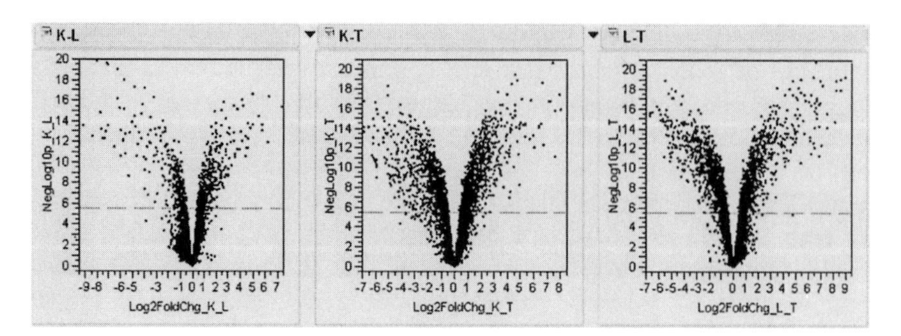

Figure 3. Volcano plots of significant differences between tissues from Model 2b when mouse is a random effect. Y axis is negative log10 based p value and x axis is the log2 based fold change

Next, we fit a model of the $\log_2$ ratios (Model 4a). Again, we take the mouse effect as a fixed effect at first, and we also model heterogeneous residual variances for three tissues independently. The distribution of the residual variance component for each tissue is displayed in Figure 4. They are close in magnitude for the three tissues. Kidney has the largest variance and testis has the smallest.

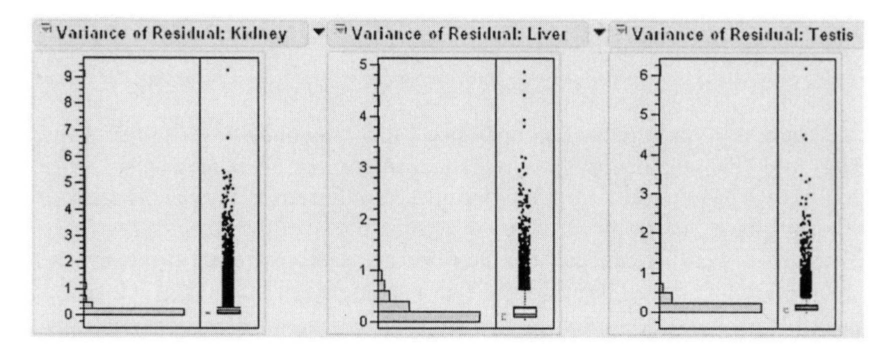

Figure 4. Variance component of Model 4a when mouse is a fixed effect

Examining the *F*-statistics of the mouse by tissue interaction for each tissue, we find 426 genes with a significant mouse effect for kidney, and compared with the Pritchard *et al.* [2001] results, we find about half of their top significant genes in our list. We also find 26 genes for liver and 94 genes for testis with overlaps of 16 for liver and 15 for testis. The overlaps are much bigger than the previous comparison since the log ratio is used as a response in this model. Table 1 also lists the top 20 genes for each tissue for this model. When comparing differential expression across tissues, over half of the genes have at least one significant difference between tissues.

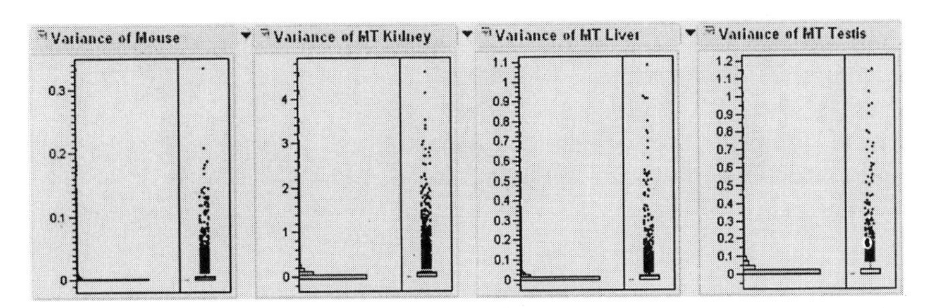

Figure 5a. Variance component of Model 4b when mouse is a random effect

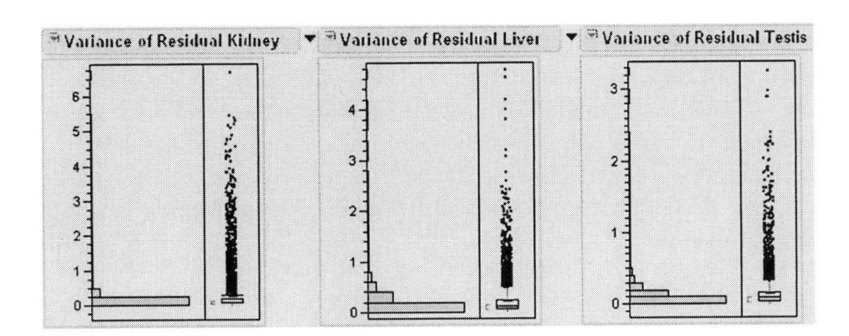

Figure 5b. Variance component of Model 4b when mouse is a random effect

Finally, we refit the above model with mouse as random effect (Model 4b). Figure 5 displays the variance component estimates. It has very similar pattern to that of Model 2b. Residual variance (array variance) is the largest, and mouse variance is the smallest. Figure 6 displays the volcano plots of the significant differences between tissues. Nearly 2000 genes are above the Bonferroni criteria for at least one tissue pair comparison.

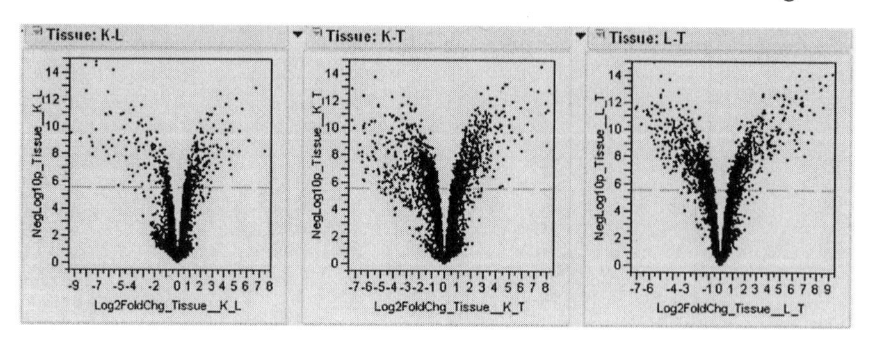

Figure 6. Volcano plots of significant differences between tissues from Model 4b when mouse is a random effect. Y axis is negative log10 based p value and x axis is the log2 based fold change.

Both approaches (log intensity and log ratio) provide similar distributional characteristics for variance components. Array always has the largest variance. When comparing differential expression between tissues, we can infer from Figures 3 and 6 that Model 2b is more sensitive than Model 4b to detect significantly differently expressed genes among tissues. Similar but smaller differences are also found between Model 2a and 4a (results not shown).

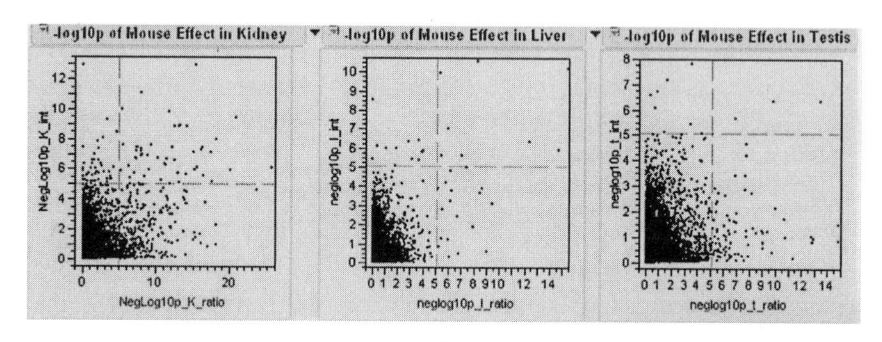

Figure 7. -log10 (p) of mouse effect for intensity model 2a (y axis) and ratio model 4a(x axis). The lines indicate the Bonferroni criterion.

To compare overall significance results, Figure 7 plots the negative $\log_{10}$ p-values for the partitioned mouse-by-tissue effects from Models 2a and 4a against each other and Figure 8 plots the negative $\log_{10}$ p-values for the tissue differences from Models 2b and 4b against each other. (The latter are the same values from the y-axes of Figures 3 and 6.) Although many more genes are significant in the tissue comparisons, it appears that Model 4a has more power than 2a to detect significant mouse-within-tissue effects (at least for kidney) whereas 2b has more power than 4b to detect significant differences between tissues (averaging across mouse effects). Note also that in Figure 7 there are several genes for which the two models differ substantially in their assessment of significance, but in Figure 8 the two models generally agree; i.e. the correlation between negative $\log_{10}$ p-values is much higher. This can likely be attributed to the different residual error variance structures of the two models, because mouse has the relatively small variation and its effect is more sensitive to the change of residual structures.

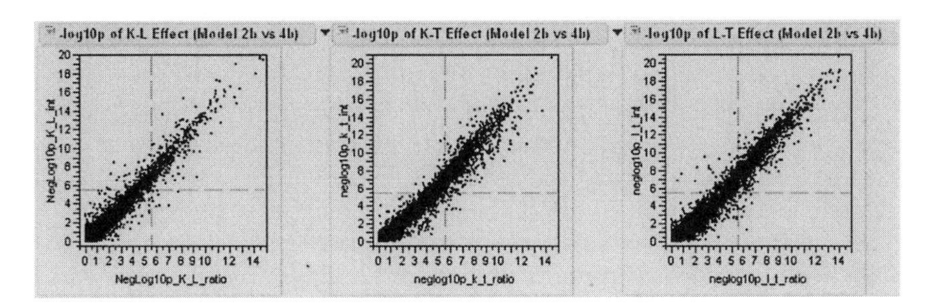

Figure 8. -log10 (p) of tissue effect for intensity model 2b (y axis) and ratio model 4b(x axis). The lines indicate the Bonferroni criterion.

4. DISCUSSION

By estimating the variance components in this experiment, we noticed one of the dominant sources of variability in cDNA microarray data is array-array variation. It is much bigger than other variance components such as mouse. This can help in experimental design questions (see Cui and Churchill [2003] in this volume.) The mixed model provides an excellent framework for rigorous prospective selection of experimental designs.

Taking mouse as a fixed effect or a random effect gave similar estimates of other variance components. When mouse is a fixed effect, we do not estimate its variance directly, but rather assess its statistical significance using F-statistics. (The fixed-effect-based F-statistic can also be used to assess the significance of the random effect variance component.) When mouse is a random effect, we can directly estimate its variance and generalize our inference to the mouse population rather than just to the observed mice. As a penalty for this additional generality, the p-values for tissue and other fixed effects will tend to be larger than those in a fixed mouse effect model.

The results in Figures 7 and 8 provide a "mixed" message about detecting significant model effects. There are some possible reasons to explain these differences as well as the lack of overlap between the top 20 gene lists generated from Models 2a and 4a and the updated list from Pritchard *et al.* [2001]. First, modeling intensities directly with the reference sample as a new treatment level is a different and more complex model on a dataset two times larger. Second, Model 2 does not include heterogeneous residual error variances, and this can result in substantial differences in how different observations are weighted. Third, different normalization procedures can contribute significant difference in the outcome. We are conducting additional studies to more thoroughly investigate the differences noticed here.

From a practical point of view, what is the best model to use? The answer depends on the study objectives and the experiment design. If we are interested in testing mouse effects as in the original study, we should consider mouse as a fixed effect. If our main interest is to assess tissue effects, a random mouse effect model is more generalizable.

Should we model intensities or ratios? When the effect of interest has only two levels (e.g. treatment and control) and they are hybridized on the two channels of the same array, there is no statistical difference in using intensities and ratios, analogous to the classical paired *t*-test. However, intensity modeling is more flexible in handling more treatment combinations in complex designs like incomplete blocks and loops, as well as other covariates. For the common reference design like this mouse normal data, the intensity model can gain power through the recovery of inter-block information using the common reference in the test of tissue effect, at a price of including array variation in the model. On the other hand, when array-array variability is large enough to offset the gain of the sample size from using the common reference in the model, one may be better off modeling ratios since the array effect is cancelled out by forming the ratio and ratios are simpler to work with. Conclusions from the two approaches generally agree, although there are cases where they can produce dramatically different results. We're pursuing further investigations in hopes of providing more specific guidelines in the future.

5. CONCLUSIONS

A linear mixed modeling approach to $\log_2$ cDNA microarray measurements is a powerful method for simultaneously assessing both variability and systematic effects. For the normal mouse data, it provides a fairly comprehensive picture of the magnitudes and interrelationships of the known sources of variability. Using $\log_2$ intensities or ratios as the response variable provides a similar pattern in variance components estimates, but some differences in the assessment of differential expression. The entire framework is useful as a foundational methodology for transcriptomic knowledge discovery from designed array experiments.

REFERENCES

Chu, T., Weir, B., and Wolfinger, R. (2002). A systematic statistical linear modeling approach to oligonucleotide array experiments. *Math. Biosci.* 176, 35-51.

Cui, X. and Churchill, G. (2003). How many mice and how many arrays? Replication in mouse cDNA microarray experiment. In S. Lin, K. Johnson (Eds), *Methods of Microarray Data Analysis III*. Boston: Kluwer.

Kerr, K., and Churchill, G. (2001). Experimental design for gene expression microarrays. *Biostatistics*. 2: 183-201.

Pritchard, C.C., Hsu, L., Delrow, J. and Nelson, P.S. (2001). Project normal: defining normal variance in mouse gene expression. *PNAS*. 98, 13266-13271. Updated data and results are at http://www.pedb.org.

Wolfinger, R. D., Gibson, G., Wolfinger, E. D., Bennett, L., Hamadeh, H., Bushel, P., Afshari, C., and Paules, R. S. (2001). Assessing gene significance from cDNA microarray expression data via mixed model. *J. Comp. Bio.*, 8, 625-637.

9

HOW MANY MICE AND HOW MANY ARRAYS? REPLICATION IN MOUSE cDNA MICROARRAY EXPERIMENTS

Xiangqin Cui and Gary A. Churchill
The Jackson Laboratory, Bar Harbor, Maine

Abstract: Biological and technical variances were estimated from the Project Normal data using the mixed model analysis of variance. The technical variance is larger than the biological variance in most genes. In experiments for detecting treatment effects using a reference design, increasing the number of mice per treatment is more effective than pooling mice or increasing the number of arrays per mouse. For a given number of arrays, more mice per treatment with fewer arrays per mouse are more powerful than fewer mice per treatment with more arrays per mouse. A formula is provided for computing the optimum number of arrays per mouse to minimize the total cost of the experiment.

Keywords: cDNA microarray, mixed model, variance components, experimental design, reference design, replication

1. INTRODUCTION

Complementary DNA (cDNA) microarrays are widely used in gene expression profiling. This complex technology involves many steps. Each step can introduce variation (technical variation), which accumulates in the final observations. Some of the systematic variation can be minimized by data transformation and normalization [Cui and Churchill, 2002; Quackenbush, 2001]. However the intrinsic variation from each step cannot be completely eliminated. Therefore, it is desirable to estimate these variance components from data and to use them to improve the statistical inference.

Biological variation is another source of variation. In order to make general claims about a treatment effect, multiple experimental units (biological replicates) from the population should be assayed. Otherwise, conclusions about the treatment effect will be restricted to the samples tested [Churchill, 2002; Cui and Churchill, 2003].

Variance components from each source can be estimated by modeling microarray data using the mixed model analysis of variance (ANOVA). Unlike fixed effect models, which treat the effects of factors as if they would be repeated exactly if the experiment were to be repeated, mixed models treat some factors, such as the array effect, as random samples from a population. In other words, we assume that if the experiment were to be repeated, the same effects would not be exactly reproduced, but that similar effects would be drawn from a hypothetical population of effects. The variation of these random factors is considered when inferences are made about the treatment effects [Littell *et al.*, 1996; Searle *et al.*, 1992; Witkovsky, 2002]. Therefore, the mixed model results are more general and reproducible.

Knowledge about variance components also provides a basis for making experimental design decisions regarding the allocation of resources [Churchill, 2002; Yang and Speed, 2002]. Replication in a microarray experiment can be present at any levels. For example, multiple samples per treatment (sample level), multiple RNA extractions per sample (RNA level), multiple labelling reactions per RNA source (labelling level), Multiple arrays per label (array level) and multiple spots per gene on each array (spot level). Replication at levels that have large variance components can significantly improve the overall sensitivity of the experiment. Current literature on microarray replication is mainly at the array level [Pan *et al.*, 2002; Wolfinger *et al.*, 2001; Zien *et al.*, 2002] and spot level [Lee *et al.*, 2000], with little attention given to biological sample level. Here we use a linear mixed model ANOVA to estimate the variance components at the mouse and the array levels and explore the implications of these variances in the allocation of biological and array replication.

2. MATERIALS AND METHODS

2.1 Data pre-process

The corrected Project Normal microarray data (text files) were downloaded from http://www.camda.duke.edu/camda02/contest.asp. Only the foreground signals were used. The background signals and flags were ignored. All genes including blanks (5776 spots on each array) are included in the analysis. The data were $\log_2$- and intensity-lowess- transformed and then normalized by subtracting the channel mean from each signal, which is the same as fitting a normalization model [Wolfinger *et al.*, 2001] or fitting some global factors for normalization in global ANOVA models [Kerr *et al.*, 2000].

2.2 Model

The following mixed linear model was fitted to each gene in each organ to estimate variance components,

$$Y_{ij} = \mu + A_i + D_j + M_k + R_h + \varepsilon_{ij}, \quad (1)$$

where μ is the gene mean; A_i (i = 1 ... 24) is array effect; D_j (j = 1, 2) is the dye effect. M_k (k = 1 ... 6) is the effect of individual mouse, where the value of k is determined by the array and dye combination (i,j). R_h (h = 1, 2) is an indicator of reference (h = 1) versus tissue sample (h = 2), which is again determined by the combination of array and dye. ε_{ij} is the residual measurement error. D_j and R_h are fixed effects. A_i, M_k, and ε_{ij} are random effects with assumed normal distributions $N(0, \sigma_A^2)$, $N(0, \sigma_M^2)$, and $N(0, \sigma_\varepsilon^2)$, respectively. σ_A^2, σ_M^2, and σ_ε^2 were estimated using the restricted maximum likelihood (REML) method [Searle *et al.*, 1992; Wikovsky, 2002].

A bigger model that includes the organ effect,

$$Y_{ij} = \mu + A_i + D_j + O_n + M_k + R_h + \varepsilon_{ij}, \quad (2)$$

is fitted to the combined data of all three organs. The organ effect O_n (n = 1, 2, 3) is a fixed effect with three levels representing kidney, liver, and testis. The number of levels in array effect A_i becomes 72 (i = 1 ... 72) in this model. The remaining terms are the same as in equation (1).

2.3 Proportion of detectable genes

The power calculation for testing the treatment effect in each gene follows Wolfinger *et al.* [2001]. The error variance (EV) for the treatment effect was calculated using the formula $EV = \sigma_M^2 / m + \sigma_\varepsilon^2 /(mn)$, where m and n are the number of mice per treatment and number of array pairs per mouse, respectively. The effectiveness of the whole experiment is represented by the proportion of detectable genes, where a gene is detectable if we have at least 50% power to detect differential expression at a given significance level and fold change.

3. RESULTS

3.1 Variance components

The variance components from the random effects - mouse (σ_M^2), array (σ_A^2), and measurement (σ_ε^2) - were estimated from the corrected Project Normal data set for each gene in each organ using the mixed model in equation (1). The mean and median of each variance component for each organ across all genes are shown in Table 1. In general, the array variance is the largest component and it is more than 10 times larger than the mouse and the measurement variances. The mouse variance is the smallest for both mean and median. Among the three organs, liver has the smallest mouse variances.

Table 1. Mean and median of variance components. The degrees of freedom for estimating variance components from mouse, array, and measurement in each gene are 5, 18, and 22, respectively.

	variance components	kidney	liver	testis
Mean	mouse	0.0252	0. 0092	0. 0126
	array	0. 3221	0. 2957	0. 3068
	measurement	0. 0250	0. 0308	0. 0244
Median	mouse	0. 0090	0. 0014	0. 0046
	array	0. 2168	0. 1848	0. 2075
	measurement	0. 0168	0. 0178	0.0154

The distributions of variance components across genes in each organ are shown in Figure 1. The histogram of the mouse variance components for individual tissue shows a characteristic bimodal pattern. The majority of the genes have small mouse variance in all three organs. Only a small portion of genes have larger mouse variances. For example, 18%, 5%, and 6% of the genes have mouse variance over 0.04 in kidney, liver and testis, respectively. This suggests that the majority of the genes are under tight control and a small portion of the genes are less tightly controlled. In addition, the identities of the less tightly controlled genes are different in the three organs. Among all genes that have mouse variance over 0.04, only 18% are common among all three organs. The low overlapping of the less tightly controlled genes could cause the mouse variances to average out when the data from all three organs were combined for variance component analysis. That is exactly what we saw when we fitted a bigger ANOVA model (equation 2) to the combined data with the organ effect accounted for. The "shoulder" in the mouse variances of kidney and testis disappeared from the mouse variance of the combined data (Figure 1D).

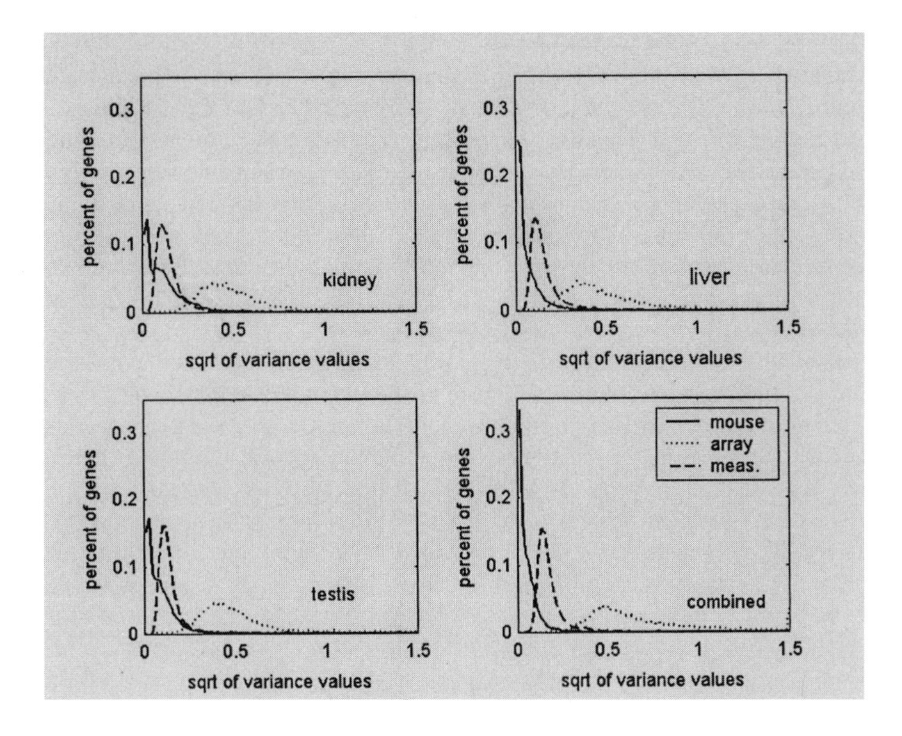

Figure 1. Distributions of the estimated variance components. Smoothed histograms of standard deviation instead of the variance are plotted for clarity.

For the purpose of making comparisons, variance components from two other experiments were estimated and are shown in Figure 2. The Gallstone experiment [Henning Wittenburg and Beverly Paigen, personal communication] looked at diet and strain effects on RNA expression in mouse liver. Two diets and three strains were assayed in a 2 x 3 factorial design with two mice for each diet and strain combination. The experiment used direct comparisons among samples on a total of 28 arrays. Each array has two adjacent spots for each gene. The Brain Cortex experiment [http://pga.tigr.org/MouseText.shtml] looked at the variation of mRNA expression of the brain cortex in mice using arrays with duplicated spots. In this case the duplicated spots were dispersed across the array. Because these data sets have duplicated spots, there are four measurements obtained for each clone, two in the red channel and two in the green channel. Measurements obtained on the same spot (one red and one green) will be correlated because they share common variation in the spot size. Measurements obtained in the same color (both red or both green) will be correlated because they share variation through a common labeling reaction. Therefore, additional random factors for spot and labeling effects were estimated using the mixed model.

The estimated mouse variance components in these two data sets are comparable with those estimated from the Project Normal data set (Figure 1). Some of the technical variance components (array, spot and measurement) are larger in the Brain Cortex experiment than in the Gallstone experiment. One possible explanation for this difference is that the dispersal of replicated spots in the Brain Cortex experiment will pick up spatial variation on the array. In addition, the difference may reflect different levels of control over spot size and morphology or hybridization quality.

Including data from all three organs in the bigger model (2) allows us to estimate the variance components from the combined data and enables us to find genes that are differentially expressed among the three organs. In this case the three organs represent three different "treatments" and finding genes that are differentially expressed could be regarded as one goal of the experiment. At significance level of 0.05 after Bonferroni correction (a stringent adjustment for multiple testing), 2279 genes were identified as differentially expressed among kidney, liver, and testis. It is not surprising to see so many differentially expressed genes given that these are three very different organs. We did not try to characterize these genes any further.

3.2 Power for detecting treatment effects

If the goal of the experiment is to detect treatment effects, statistical inference should be based on the total variance including the biological variance. This will typically be a weighted average of the variance components and the relative weighting will depend on the design of the experiment. The error variance (*EV*) for treatment effect is a combination of all the variance components that are nested under treatment.

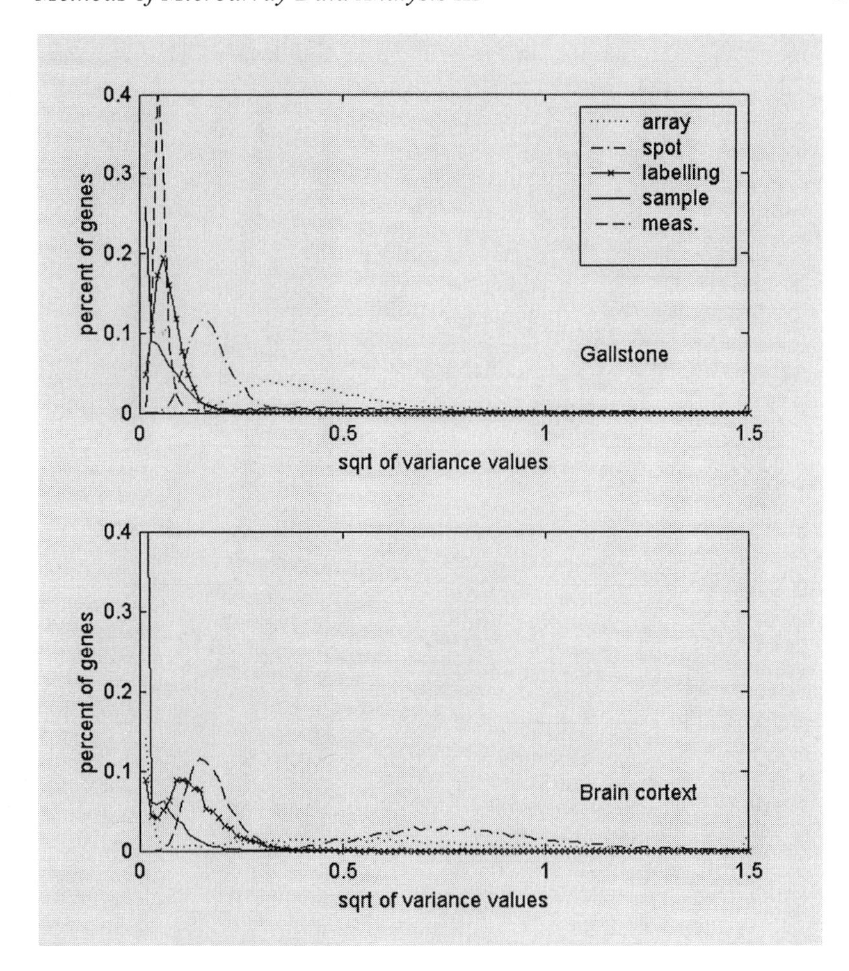

Figure 2. Variance components from mouse Gallstone and Brain Cortex data sets. The smoothed histograms of standard deviations instead of the variances are plotted for clarity.

In a simple reference design for comparing two treatments, TrtA and TrtB (Figure 3), the EV for treatment effect can be computed by

$$EV = \frac{\sigma_M^2}{m} + \frac{\sigma_\varepsilon^2}{mn} \tag{3}$$

with m as number of mice per treatment and n as pairs of dye-swap arrays per mouse. This equation is obtained using the within-array information only as in tests based on log ratios. The mixed model can extract between-array information and combine it with the within-array information [Littell *et al.*, 1996]. The *EV* from the mixed model (equation 1) will be

$$EV = \frac{\sigma_M^2}{m} + \frac{\sigma_\varepsilon^2}{mn} - \frac{\sigma_\varepsilon^4}{2mn(\sigma_A^2 + \sigma_\varepsilon^2)} \quad (4)$$

We have seen that array variance σ_A^2 dominates, thus the correction factor will generally be quite small and we use equation (3) throughout the rest of this paper. Note that the array variance is not included here. This is a consequence of the pairing in two-dye microarray experiments. In one-color systems, array variance will be a major component of *EV*.

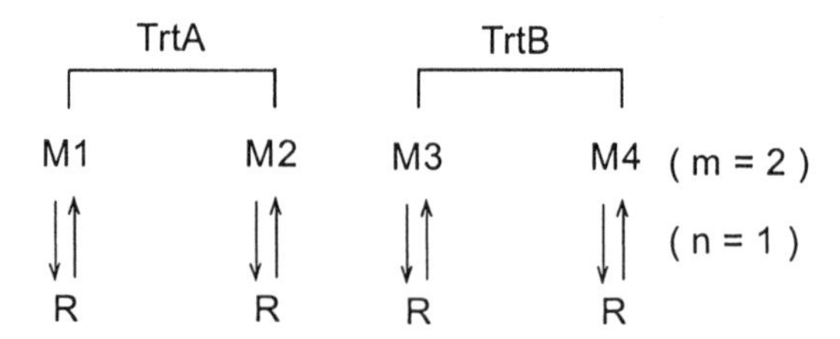

Figure 3. Reference design for testing treatment effect TrtA versus TrtB. Arrows represent arrays with head as Cy3 and tail as Cy5. M=mouse; R=reference; m=number of mice per treatment; n=number of array pairs per mouse.

A small error variance is desirable in order to increase the power of statistical tests (*t* or *F*) for treatment effects. From equation (3) we can see that this can be achieved by increasing the number of mice per treatment, *m*, to reduce both components proportionally. However, increasing *m* may mean a substantial increase in cost when mice are expensive. Therefore, it may be desirable to simply increase the number of array pairs per mouse, *n*. However, this strategy will only reduce the technical component of the variance, therefore, it is most effective when measurement variance is larger than the mouse variance.

Figure 4 shows the proportion of genes in which a two-fold difference between two treatments can be detected with at least 50% power at a significance level of 0.05 after Bonferroni correction at various combinations of *m* and *n* in the three organs. The *m=2* (2 mice per treatment) case does not show any power, while the 4, 6 and 8 mice per treatment cases show substantially increased power. The lack of power in the *m=2* case is mainly due to the small degrees of freedom (*df*) for estimating the variance in the t test. Since the right side of equation (3) is the sum of two variance components, we use the smaller df as a conservative approximation to the df of the error variance. Using this method, the df of the $m=2$ case is 2 while those of the other cases are 6, 10, and 14. At least 5 degrees of freedom are generally recommended for a t test.

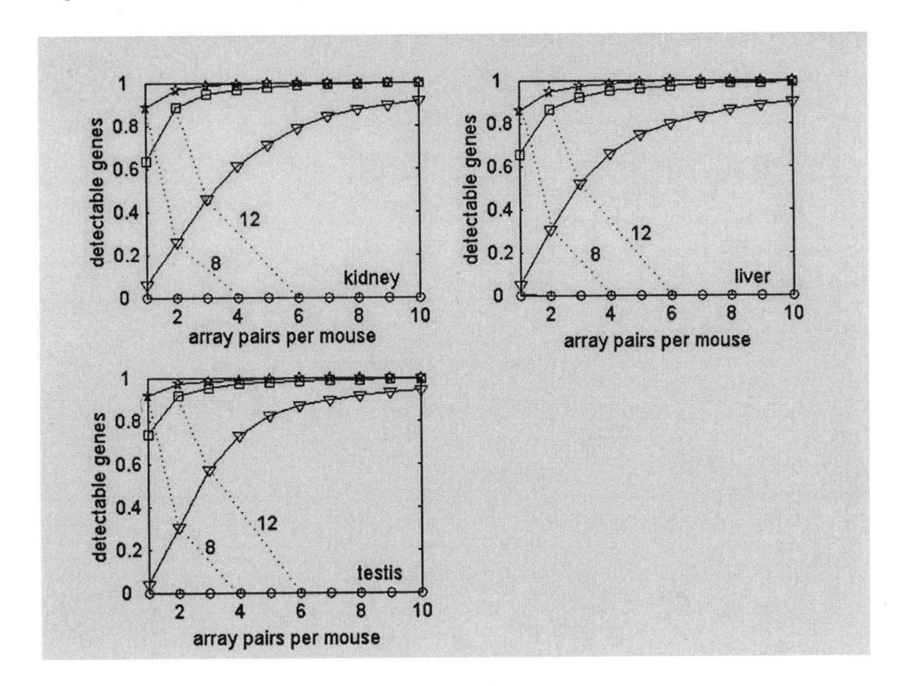

Figure 4. Power for detecting 2-fold change between two treatments at various combinations of number of mice per treatment and number of array pairs per mouse. Circle, triangle, square, and star represent 2, 4, 6, and 8 mice per treatment. Dotted lines represent the same number of array pairs (8 or 12) for each treatment. Significance level is 0.05 after Bonferroni correction.

Increasing the number of arrays per mouse generally increases the proportion of detectable genes. The increase is most obvious with small numbers of arrays (from 1 to 4 pairs). Once there are 6 pairs of arrays per mouse, there is hardly any gain from increasing arrays further. If the goal is to achieve detectability for more than 90% of the genes, at least 6 mice per treatment is recommended. For a small number of genes which have significantly larger mouse variances, such as the genes declared significant in Pritchard *et al.* [2001], increasing the number of arrays per mouse can never achieve the same precision as increasing the number of mice per treatment. For the same total number of arrays per treatment, more mice per treatment with fewer arrays per mouse will have more statistical power as shown by the dotted lines in Figure 4. Therefore, if mice are relatively inexpensive, more mice and fewer arrays per mouse is a better choice.

The percent of genes that can be detected at various fold changes is shown in Figure 5 for various combinations of arrays and mice. Increasing array number per treatment can greatly increase the detection of genes with smaller fold changes. In addition, the experiment with six mice per treatment has substantially higher power than that with four mice per treatment at small fold changes.

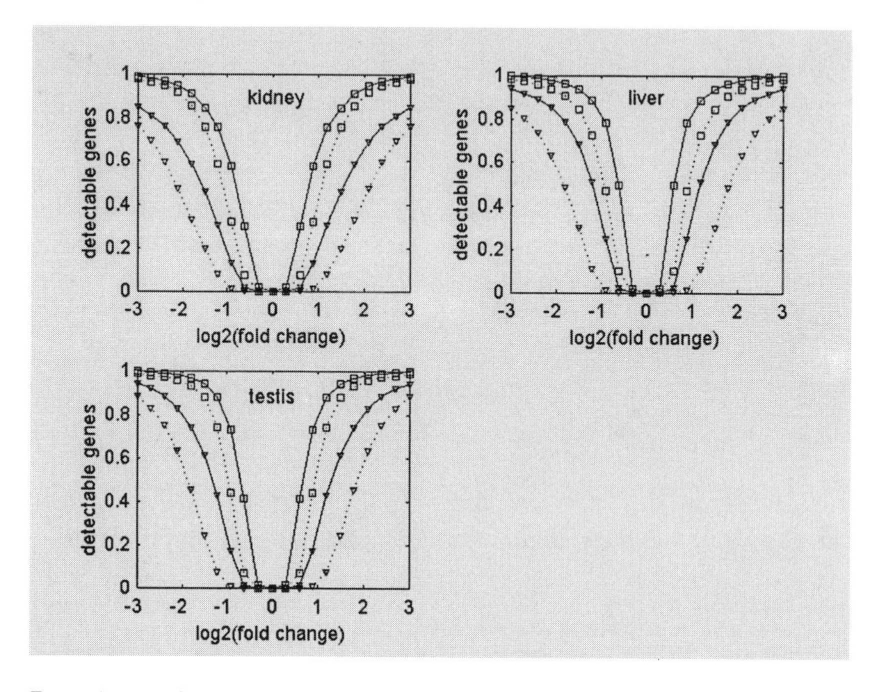

Figure 5. Power for detecting treatment effects at different fold changes. Triangles and squares represent 4 and 6 mice per treatment, respectively. Dotted and solid lines represent 1 and 3 pairs of arrays per mouse, respectively. Significance level is 0.05 after Bonferroni correction.

3.3 Optimum resource allocation

In practice, there is often a limited budget. In order to utilize the resource most effectively, we need to balance mice and arrays to minimize the cost of the whole experiment. Let C_M represent the cost of a mouse and C_A represent the cost of a pair of arrays. Suppose there are m mice per treatment and each mouse will be measured using n pairs of arrays. The total cost per treatment will be

$$Cost = m \cdot C_M + m \cdot n \cdot C_A \tag{5}$$

Combining equations (3) and (5), we can derive the optimum number of arrays per mouse [Kuehl, 2000] in order to keep the total cost minimum as

$$n = \sqrt{\frac{\sigma_\varepsilon^2}{\sigma_M^2} \cdot \frac{C_M}{C_A}} \tag{6}$$

For example, suppose that the array price is \$300 each and three different mouse strains are \$15, \$300, and \$1500 per mouse. If we use the median values of $\sigma_\varepsilon^2 / \sigma_M^2$ obtained from each organ of the Project Normal data, the optimum pairs of arrays per mouse for differently priced mouse strains can be computed using equation (6). Because all the calculations are based on reference design with dye swap (Figure 3), n is rounded up to the nearest integer (Table 2).

Table 2. Optimum pairs of arrays per mouse (N = number of pairs of arrays per mouse)

Array price (pair)	Mouse price	N (kidney)	N (liver)	N (array)
\$600	\$15	1	1	1
\$600	\$300	1	3	2
\$600	\$1500	2	6	3

If the budget is fixed, equation (6) can be plugged into equation (5); the resulting m and n values will give a minimum EV for the fixed cost. If there is a certain EV to achieve, equation (6) can be plugged into equation (3) to find an m and n that will result in minimum cost for the desired EV.

When there are r replicated spots for each clone on each array, additional variance components for labeling (σ_L^2) and spot can be fitted to the data to capture the covariance shared among observations within a spot and a labeling reaction as shown in Figure 2. The EV of the treatment mean can be approximated as

$$EV = \frac{\sigma_M^2}{m} + \frac{\sigma_L^2}{mn} + \frac{\sigma_\varepsilon^2}{mnr} \tag{7}$$

and the optimum number of dye-swap pairs per mouse will be

$$n = \sqrt{\frac{r\sigma_L^2 + \sigma_\varepsilon^2}{r\sigma_M^2} \cdot \frac{C_M}{C_A}} \tag{8}$$

3.4 Pooling mice

Pooling mice is another possible strategy to increase the precision of treatment tests in this experiment. It reduces *EV* by reducing the mouse variance component. The between-pool variance for a pool size of k mice will be approximately

$$\sigma_{pool}^2 = \frac{1}{k^\alpha}\sigma_M^2 \tag{9}$$

for some constant $0 < \alpha < 1$, which is related to the pooling procedure. In the case of $\alpha = 0$, pooling will have no effect. In the case of $\alpha = 1$, the mouse variance is reduced in direct proportion to the pool size k. Unfortunately, there is no information about α in the Project Normal data set. Therefore, we cannot estimate how effective pooling will be in this experiment. Suppose that we can reach the maximum effect of pooling, $\alpha = 1$, the effect of pooling 3 mice in each sample is shown in Figure 6. The power for detecting treatment effect increases slightly in kidney and testis, even less in liver based on the variance components estimated from the Project Normal data set. However, pooling biological sample will be more effective in experiments where biological variance is the major source of variance.

Due to the presence of α in equation (9), for a fixed total number of mice, more pools with fewer mice in each pool will result in a smaller *EV*, given that a fixed number of arrays for each treatment will be used. For example, 10 pools with 4 mice in each pool are better than 4 pools with 10 mice per pool.

4. DISCUSSION

In this paper we estimated the biological and technical variance components from the Project Normal data and found that the technical variance is the major component in the treatment error variance. Therefore, reducing the technical variance will be effective in increasing the power of the test for treatment effect. One way to reduce the technical variance is through increasing the number of arrays per mouse as discussed in the result section. Another approach, not considered here, is to use more efficient experimental designs, such as loop design [Kerr and Churchill, 2001a; Kerr

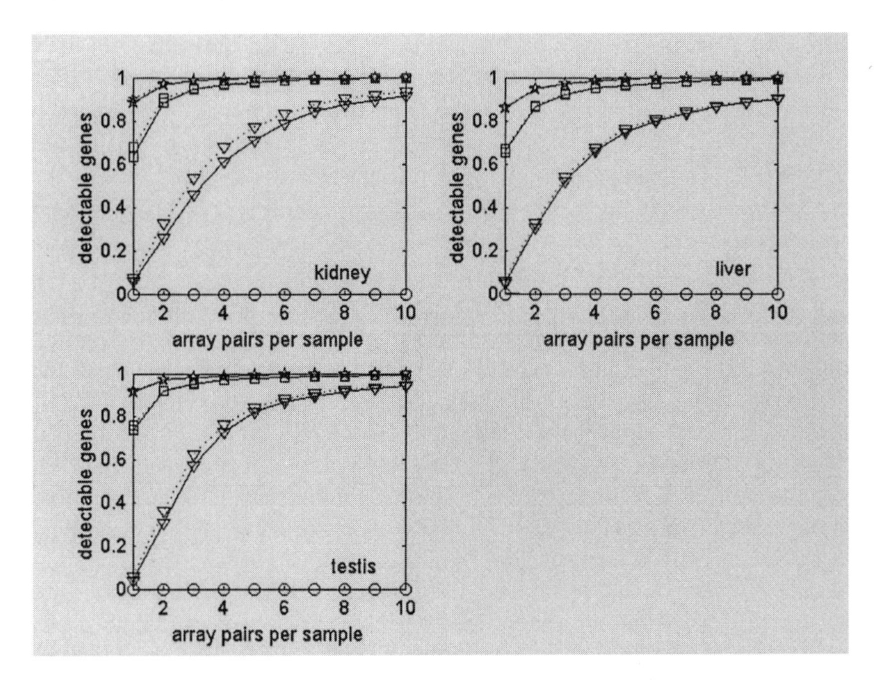

Figure 6. Effect of pooling mice in increasing the power of detecting treatment effect. Circle, triangle, square, and star lines represent 2, 4, 6, and 8 mice (solid) or pools (dotted). Each pool has 3 mice. Significance level is 0.05 after Bonferroni correction.

and Churchill, 2001b] and block design [Simon *et al.*, 2002]. No reference sample is used in these designs and the treatment samples are compared directly. Direct comparisons have higher efficiency than indirect comparisons through the reference sample, because no measurement is wasted on the reference sample [Kerr and Churchill, 2001a; Yang and Speed, 2002]. For comparing two samples, the technical variance from a direct comparison is ¼ of that from an indirect comparison through a reference sample using the same number of arrays. For more than two samples, the reduction of technical variance depends on the positions of comparing samples in the design and it becomes more complicated to calculate when complicated loop or block designs are used [Yang and Speed, 2002; Churchill, 2002].

The power calculations in this paper are based on the assumption that each gene has unique variance components, which is a relatively unstable method in experiments with a limited number of data points per gene. In those experiments, we could assume that the variance components are the same across all genes; in this case the proportion of detectable genes will be either 0% or 100% because the test statistic (t or F) value will only depend on the magnitude of the fold change, not the variance components from each gene. However, this type of test is subject to bias if the data have not been properly normalized. If we combine the above two situations and assume that each variance component for each gene is the combination of common variation for all genes and some variation from individual genes [Baldi and Long, 2001; Cui and Churchill, 2003; Lönnstedt and Speed, 2002], the proportion of detectable genes will again be a useful concept. Combining information about variance components across the genes is a potentially powerful approach that we are currently investigating.

Multiple testing adjustments are usually applied to control the false positive errors in microarray experiments when thousands of genes are tested one at a time as in this experiment. In this experiment, we used the Bonferroni correction, which is a stringent family-wise error rate (FWER) correct. The 0.05 significance level after Bonferroni correction means that we expect a probability of 0.05 to have one or more errors in the whole list of the identified significant genes. There are other less stringent multiple testing adjustment methods, such as FDR (false discovery rate) adjustments, which controls the percent of genes in the declared significant gene list that are false [Benjamini and Hochberg, 1995; Storey, 2002]. These methods are appropriate for exploratory experiments in which a list of candidate genes will be confirmed using other technologies. When less stringent multiple test adjustments are used, the power of the test in Figures 4, 5, and 6 will all increase.

All calculations of power and resource allocation in this paper are based on the estimation of the variance components, which relies on replication. In the Project Normal data set, mouse and array are replicated; therefore, the variance introduced by these two factors can be estimated. Similarly, variance components from any other factor, such as RNA extraction, labeling or spot, can be estimated as long as there is replication at an appropriate level of the experimental design. Estimation of all the possible variance components from microarray technology can help to determine where the largest variances come from. These steps are targets for microarray technology improvement. The estimation of variances becomes even more important when new microarray platforms, new techniques, or new facilities are implemented. For large and complicated experiments, pilot studies are recommended for estimating the biological and technical variances in order to customize the power and cost calculation in the design stage.

5. CONCLUSION

We estimated the biological and technical variance components from the Project Normal data set using a mixed model ANOVA. Comparison among the estimated variance components revealed that the technical variance is larger than the biological variance for most of the genes. To detect treatment effects using a reference design in experiments with similar variance component values, reducing the biological variance by pooling mice will not be as effective as increasing the replication of arrays to reduce the technical variance. For a fixed number of arrays per treatment, designs with more mice per treatment and fewer arrays per mouse are more powerful than designs with fewer mice per treatment and more arrays per mouse.

ACKNOWLEDGMENTS

We thank Hao Wu for helping the software development and Jason Stockwell for valuable comments. We also thank Henning Wittenburg and Beverly Paigen at the Jackson Lab for the Gallstone data and John Quackenbush at TIGR for the Brain Cortex data. This research is supported by grants CA88327, HL66620, and HL55001 from the National Institute of Health.

REFERENCES

Baldi, P, and Long, AD (2001). A Bayesian framework for the analysis of microarray expression data: regularized t -test and statistical inferences of gene changes. Bioinformatics 17: 509-519.

Benjamini, Y, and Hochberg, Y (1995). Controlling the false discovery rate: A practical and powerful approach to multiple testing. J R Stat Soc B 57: 289-300.

Churchill, GA (2002). Fundamentals of experimental design for cDNA microarrays. Nat Genet 32 Suppl 2: 490-495.

Cui, X, and Churchill, GA (2002). Data transformation for cDNA microarray data. http://www.jax.org/staff/churchill/labsite/pubs/index.html.

Cui, X, and Churchill, GA (2003). Statistical tests for differential expression in cDNA microarray experiments. Genome Biol 4: 210.

Kerr, MK, and Churchill, GA (2001a). Experimental design for gene expression microarrays. Biostatistics 2: 183-201.

Kerr, MK, and Churchill, GA (2001b). Statistical design and the analysis of gene expression microarray data. Genet Res 77: 123-128.

Kerr, MK, Martin, M, and Churchill, GA (2000). Analysis of variance for gene expression microarray data. J Comput Biol 7: 819-837

Kuehl, R (2000). Designs of experiments: statistical principles of research design and analysis, 2 edn, Duxbury Press).

Lee, ML, Kuo, FC, Whitmore, GA, and Sklar, J (2000). Importance of replication in microarray gene expression studies: statistical methods and evidence from repetitive cDNA hybridizations. Proc Natl Acad Sci U S A 97: 9834-9839.

Littell, RC, Milliken, GA, Stroup, WW, and Wolfinger, RD (1996). SAS system for mixed models (Cary, NC, SAS institute Inc.,).

Lönnstedt, I, and Speed, T (2002). Replicated Microarray Data. Statistica Sinica 12: 31-46.

Pan, W, Lin, J, and Le, CT (2002). How many replicates of arrays are required to detect gene expression changes in microarray experiments? A mixture model approach. Genome Biol 3: research0022.

Pritchard, CC, Hsu, L, Delrow, J, and Nelson, PS (2001). Project normal: defining normal variance in mouse gene expression. Proc Natl Acad Sci U S A 98: 13266-13271.

Quackenbush, J (2001). Computational analysis of microarray data. Nat Rev Genet 2: 418-427.

Searle, SR, Casella, G, and McCulloch, CE (1992). Variance components, John Wiley and sons, Inc.).

Simon, R, Radmacher, MD, and Dobbin, K (2002). Designs of studies using DNA microarrays. Genet Epidemiol 23: 21-36.

Storey, J (2002). A direct approach to false discovery rates. J R Statist Soc B 64: 479-498.

Witkovsky, V (2002). MATLAB algroithm mixed.m for solving Henderson's mixed model equations. http://www.mathpreprints.com

Wolfinger, RD, Gibson, G, Wolfinger, ED, Bennett, L, Hamadeh, H, Bushel, P, Afshari, C, and Paules, RS (2001). Assessing gene significance from cDNA microarray expression data via mixed models. J Comput Biol 8: 625-637.

Yang, YH, and Speed, T (2002). Design issues for cDNA microarray experiments. Nat Rev Genet 3: 579-588.

Zien, A, Fluck, J, Zimmer, R, and Lengauer, T (2002). Microarrays: how many do you need? Proc. RECOMB'02. 321-330.

10

BAYESIAN CHARACTERIZATION OF NATURAL VARIATION IN GENE EXPRESSION

Madhuchhanda Bhattacharjee[1], Colin Pritchard[2], Mikko J. Sillanpää[1] and Elja Arjas[1]
[1]Rolf Nevanlinna Institute, University of Helsinki, Finland [2]Division of Human Biology, Fred Hutchinson Cancer Research Centre, Seattle, WA.

Abstract: For gene expression data we propose a hierarchical Bayesian method of analysis using latent variables, wherein we have combined normalization and classification in a single framework. The uncertainty associated with classification for each gene can also be estimated based on the posterior distributions of the latent variables applied. The proposed models are implemented using the MCMC algorithm.

Key words: Bayesian latent class models, gene expression data, MCMC.

1. INTRODUCTION

We present a new latent-variable-based Bayesian clustering method for classifying genes into categories of interest. The approach is integrated in the sense that normalization and classification can be carried out jointly. This is done along with estimation of uncertainty that makes it unnecessary to test a large number of hypotheses. Possible distortion in measuring the actual expression level due to factors like environmental and experimental conditions, dye, etc. are incorporated into the normalization part of the model. The observed expression is treated as a "black box" for the different effects, which are considered jointly in a nested common structure. The adjusted expression ratios are then classified into different categories of interest. The approach is very general in the sense that it is easily customizable for different needs and can be modified with the availability of additional information.

Preliminary and extended versions of the model were applied to the expression data provided by Pritchard *et al.* [2001]. The classification categories of interest are variational categories of genes in normal circumstances. Our findings support the hypothesis that, apart from the fact that there are several sources of variation affecting the observed expression of the genes, some genes by nature exhibit highly varied expression.

Several other Bayesian approaches have been presented recently for microarray data analysis [Medvedovic, 2000; Keller *et al.*, 2000; Long *et al.*, 2001; Baldi and Long 2001; Newton *et al.*, 2001; Ibrahim *et al.*, 2002; Dror *et al.*, 2002; Parmigiani *et al.*, 2002]. However, our approach differs substantially from the others in several ways. For example, we consider model-based normalization of data, whereas normalization is usually considered as a separate procedure without accounting for its effect on subsequent classification. Also, given the unusual nature of the data considered here, we have primarily considered classification with respect to variance whereas most microarray data analysis concentrates on classification with respect to the central tendency of the (log) expression ratios for a gene.

2. DATA

The data contained median foreground and background intensities for 5776 spots from experimental and reference samples taken from 3 organs of 6 mice each applied with 2 dyes and 2 replicates. This resulted in more than 1.5 million data points. Of the 5776 spots on the arrays 5401 were mouse genes. The differences of foreground and background intensities in the experimental and reference probes were treated as observed intensities. On several occasions the resulting intensities turned out to be negative. In absence of further clarification for such measured intensities, these were treated as missing data. Modeling was carried out after taking the natural logarithm of the experimental and reference probe intensity ratio.

3. PRELIMINARY MODEL (MODEL A)

Given that both the experimental and the reference sample are to represent the normal expression for that gene, the observed variations in the log ratio of the corresponding measured intensities can then be attributed to different and possibly nested sources, such as, mouse or dye or replicate. Without further subdividing the possible sources of variation, we assume that such factors affect all the genes and accordingly adjust the observed expression log ratios by an effect for each organ and each of the 24 arrays, denoted by μ_{OJ}, O = K (kidney), L (liver), and T (testis), J=1,...,24.

The adjusted data were then inspected for possible variation still remaining (if any) exhibited by the genes. It is anticipated that the genes may naturally behave differently in different organs, e.g. by varying highly in one organ but not in another. Accordingly each gene was classified independently for each organ with respect to its corresponding residual variance. The adjusted log-ratios were classified for the unknown (possibly natural) variation of the genes for each organ into three different categories. We assume three latent variance classes with (unknown) ordered variances $\sigma_1^2 > \sigma_2^2 > \sigma_3^2$. For each gene I and for each organ O, let C_{IO} indicate its variance-class membership for that organ. We assume that C_{IO} takes a value in range {1,2,3} with probabilities $\{\lambda_1, \lambda_2, \lambda_3\}$. The modeling was actually carried out using corresponding precision parameters (i.e. inverse of variance) $\tau_1 < \tau_2 < \tau_3$. It may be mentioned here that it is also possible to consider large or even an infinite number of variance classes through Bayesian infinite mixture, but for practical implementation and better interpretability we restrict our models to a finite mixture of latent classes (see Section 7 for a discussion).

Following the above notation, for the Ith gene, O^{th} organ and J^{th} array the conditional distribution of the log-ratio of intensity I_{IOJ} given the corresponding membership indicator C_{IO} is assumed to be given by

$$I_{IOJ} = \mu_{OJ} + E_{IOJ}, \text{ where } E_{IOJ} \sim N(0, 1/\tau(C_{IO})) \quad (1)$$

where the arrays are ordered as in the original data. That is, we start with four arrays from mouse 1 with the first two being the two replicates with green dye applied to the experimental sample, and so on.

In the following we adopt the notation that a vector or a collection of certain parameters/variables is represented by suppressing the subscripts. The posterior density $P(\mu,\tau,C,\lambda|I)$ is proportional to the joint density $P(\mu,\tau,C,\lambda,I)$ and by assuming suitable conditional independence properties between the parameters to hold, it can be presented in the product form $P(I \mid \mu,\tau,C)\ P(C \mid \lambda)\ P(\tau)\ P(\mu)\ P(\lambda)$. We assume vague priors for all model parameters. The array effects (μ) were assigned Normal priors, N(0,10). The precision parameters (τ) were assumed to have Gamma distributions *a priori*, $\tau_1 \sim \Gamma(1,1)$, $\tau_{j+1} = \tau_j + \eta_j$ where $\eta_j \sim \Gamma(1,1)$ and j = 1, 2. The latent class-indicators (C) were assigned Multinomial distributions with corresponding probabilities (λ) drawn from a Dirichlet distribution, D(**1**). In order to preserve compatibility, the estimation of the model parameters for all three organs was carried out simultaneously.

3.1 Model Implementation and Sensitivity Analysis

We implemented the model and performed parameter estimation using WinBUGS [Gilks *et al.*, 1994]. Missing data points were treated in the same way as parameters in our model and were completed during estimation using Bayesian data augmentation [Gelman *et al.*, 1995]. Ten thousand Markov chain Monte Carlo (MCMC) samples were drawn based on multiple (parallel) chains with additional burn-in rounds. The convergence of the chain was monitored by CODA [Best *et al.*, 1995] and by inspecting the sample paths of several of the model parameter estimates. Additional sensitivity analysis was carried out by comparing the posterior distributions arising using different choices of the hyper-parameters in the priors. Based on results of these analyses, we concluded that, due to the large sample size and well-structured data, the posterior distributions of population parameters were almost identical independently of the choice of the prior.

3.2 Model A: Results

From the posterior estimates of μ's (Figure 1), it was observed that, apart from array specific variations in the estimates, the estimates clearly depict an effect of dye on the observed log-ratio of intensities, especially in samples from kidney and testis. For these, all arrays in which the experimental sample had been treated with green dye had estimated means higher than the corresponding means resulting from treatment with red dye, which is a phenomenon commonly observed in many microarray experiments. Such an effect may be present in all genes but the magnitudes might differ from gene to gene. Arrays based on the samples from testis showed not only a dye effect in the array means but also indicated a possible mouse effect.

However, to what extent the observed mouse effect is caused by biological factors and to what extent it is affected by experimental artifacts like RNA preparation, is not identifiable from the present data.

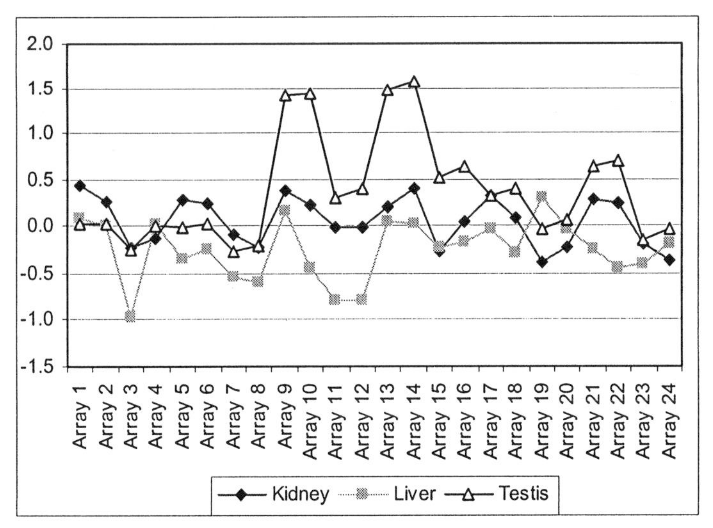

Figure 1. Plots of estimated posterior means for 24 arrays each from the three organs (kidney, liver, and testis) with Model A. Different line types are given in the bottom. Arrays are ordered according to 6 mice (M), 2 dyes (D) and 2 replicates (R) as (M1,D1,R1), (M1,D1,R2), (M1,D2,R1), (M1,D2,R2), (M2,D1,R1), ...

The posterior distributions of the three precision parameters (τ's) were quite separated from each other, depicting three distinct variance categories for the genes. Also, the estimated distributions were highly concentrated around the posterior mean, indicating un-ambiguity about the corresponding variance class.

Table 1. Posterior estimates of (common) precision parameters and gene proportions in the three precision groups (1,2,3) under Model A in the three organs.

Parameter	Group	Kidney	Liver	Testis
Precision	1	0.3	0.3	0.3
	2	2.9	2.9	2.9
	3	14.2	14.2	14.2
Percentages of genes	1	13.0	12.2	7.6
	2	40.1	40.1	41.7
	3	46.9	47.7	50.7

The estimated variance class for the residual variances for each gene (I) in each organ (O) was obtained from posterior distributions of the corresponding C_{IO}'s. It was characteristic of the results that, *a posteriori*, the genes were assigned to a variance class (within each organ) quite distinctly and for many genes the estimated membership probabilities were either near zero or near one, although the model puts no such constraints and priors were chosen to be non-informative. Also possibilities of bad mixing / non-convergence were effectively ruled out by validating the posterior estimates from several parallel chains with distinctly different initial values.

From Table 1 we observe that the smallest number of highly varying genes was observed in testis (7.6%) and the largest in kidney (13.0%).

From the estimated variance-classes, it was observed that some of the genes behave differently across organs (Table 2). Although a large proportion of the genes was identified to be less varying in all three organs, there were genes which dramatically changed the assignment of variance class from one organ to another. About 79% of genes were estimated to belong to moderate or low variance classes in all three organs. Only 1.7% of genes were estimated to have high variation in all three tissues.

Table 2. Cross tabulation of genes according to their estimated variation groups (1: High, 2: Moderate, 3: Low) in different organs (K: kidney, L: liver, T: testis) under Model A (in %).

% of genes		(T,1)	(T,2)	(T,3)	(T,1)	(T,2)	(T,3)	(T,1)	(T,2)	(T,3)	Total
(K,1)	(L,1)	1.7	6.0	0.4							8.1
	(L,2)	0.7	2.7	0.9							4.3
	(L,3)	0.2	0.2	0.2							0.6
(K,2)	(L,1)				0.7	1.5	0.5				2.8
	(L,2)				2.6	11.4	7.7				21.8
	(L,3)				0.6	8.0	7.0				15.6
(K,3)	(L,1)							0.6	0.4	0.3	1.3
	(L,2)							0.6	6.0	7.5	14.1
	(L,3)							0.0	5.3	26.2	31.5
Total		2.6	9.0	1.4	4.0	20.9	15.2	1.1	11.8	34.0	100.0

4. EXTENDED MODEL (MODEL B)

It is expected that some genes may be expressed differently in one organ compared to their average expression in all three organs. This indicates that for these genes, for a particular organ, the observed log-ratio-of-intensities can be away from the expected zero value. That is, they can have a higher or lower expression in the experimental sample in one particular organ when compared to the expression in the reference sample. Since the reference samples were taken from all three tissues, it can then be expected that for the same genes in one or both of the remaining organs the log-ratio-of-intensities would behave in the opposite way than in the first organ.

In order to account for such between-organ differences in gene expression we modeled the log-ratio of intensities for the genes in the following manner. The extended model continued to have array mean effects across all genes (i.e. μ's) as in Model A. Additionally each gene (I) was classified independently in each organ (O) as belonging to one of three possible expression groups (D_{IO}). Accordingly for each gene, an expression level ($\theta(D_{IO})$) was added to the corresponding array mean effect (μ_{OJ}) to explain the expected log-ratio of intensities of that gene.

As in Model A, each gene was classified independently for each organ with respect to its residual variance (C_{IO}).

In summary, the conditional distribution of the log-ratio-of-intensity I_{IOJ},, given C_{IO} and D_{IO}, was assumed to be given by (with I, O, J as before),

$$I_{IOJ} = \mu_{OJ} + \theta(D_{IO}) + E_{IOJ}\text{ , where } E_{IOJ} \sim N(\,0,\, 1/\tau(C_{IO})) \quad (2)$$

The prior assumptions on the original parameters were kept unchanged. In addition, the θ's were assigned Normal priors (on appropriate ranges) with $\theta_1 < \theta_2 = 0 < \theta_3$ and, similar to C_{IO}'s, the latent variables D_{IO} were assumed to be *a priori* Multinomial with corresponding probabilities (λ_D) drawn from a Dirichlet distribution. The posterior density $P(\mu,\tau, C, \lambda_C , D, \lambda_D |I)$ is then defined in a manner similar to the previous model.

4.1 Model B: Results

According to the extended model a gene in organ o and array J can now be assigned to one of three different means parameterized by $\mu_{OJ} + \theta(D_{IO})$. The estimated values of $\mu_{OJ} + \theta(D_{IO})$ with $D_{IO} = 2$ were comparable to the average array effects μ_{OJ} obtained under the previous model. On the other hand, $\mu_{OJ} + \theta(D_{IO})$ with $D_{IO} = 1$ correspond to a lower expression class, and those with $D_{IO} = 3$ to a higher expression category. Plots of posterior estimates of $\mu_{OJ} + \theta(D_{IO})$ clearly depict the distinct nature of the three expression levels. Those for kidney samples have been presented in Figure 2. Similar observations were made for liver and testis samples.

The estimated variance-classes improved in the sense that each of the estimated precision parameters (τ) under the new model is higher than that under the preliminary model. In other words, allowing for gene specific adjustment of expression levels in the extended model explained more variation in the observed expressions thereby leading to reduced residual variances (Table 3).

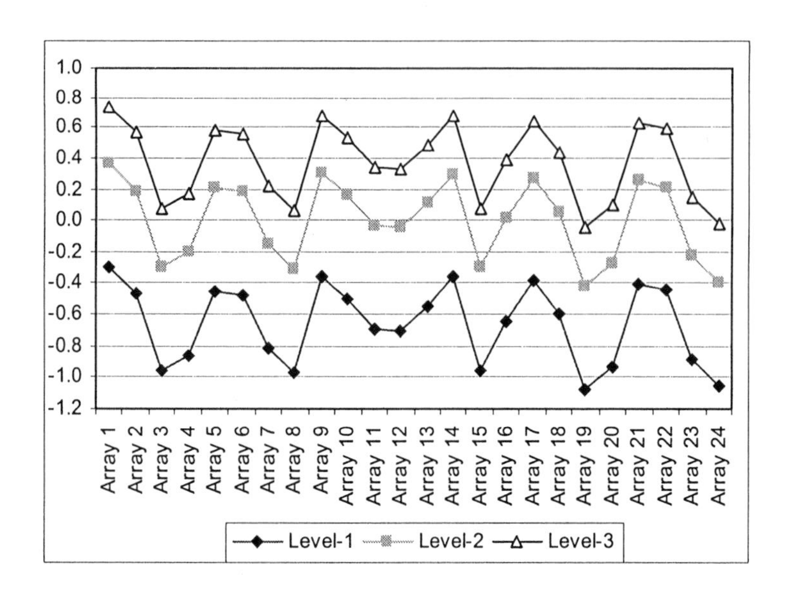

Figure 2. Kidney data: Plots of posterior estimates of $\mu_{OJ} + \theta(D_{IO})$ for O = K, J = 1, ..., 24 and three different expression levels corresponding to D_{IO} = 1 (lower expression), 2 (average expression) and 3 (higher expression) with array ordering as in Figure 1.

Moreover, even though the new variance classes gave rise to narrower distributions for the expressions, the estimated proportions of genes in the lowest variance-class still increased from Model A to Model B. Also the number of genes in the highest variance class was reduced compared to Model A (Table 3).

Table 3. Posterior estimates of (common) precision parameters and gene proportions in the three precision groups (1,2,3) under Model B in the three organs.

Parameter	Group	Kidney	Liver	Testis
Precision	1	0.5	0.5	0.5
	2	4.7	4.7	4.7
	3	19.0	19.0	19.0
Percentages of genes	1	11.1	8.7	5.0
	2	34.3	42.7	36.7
	3	54.7	48.6	58.3

Under Model B, more genes were estimated to have moderate or low variation in all three organs, compared to A (Table 4). For some genes, the estimated variance classes still varied across the three organs, although the number of such genes was smaller than in the previous model.

Table 4. Cross tabulation of genes according to their estimated variation groups (1: High, 2: Moderate, 3: Low) in different organs (K: kidney, L: liver, T: testis) under Model B (in %).

% of genes		(T,1)	(T,2)	(T,3)	(T,1)	(T,2)	(T,3)	(T,1)	(T,2)	(T,3)	Total
(K,1)	(L,1)	1.4	2.7	1.2							5.2
	(L,2)	0.8	3.4	0.7							4.9
	(L,3)	0.1	0.6	0.2							0.9
(K,2)	(L,1)				0.5	1.1	0.4				2.0
	(L,2)				0.8	12.7	7.7				21.2
	(L,3)				0.3	5.2	5.7				11.1
(K,3)	(L,1)							0.4	0.4	0.6	1.4
	(L,2)							0.6	4.8	11.2	16.6
	(L,3)							0.1	5.9	30.6	36.6
Total		2.3	6.6	2.1	1.6	19.0	13.7	1.1	11.1	42.5	100.0

From the estimated θ values and the estimated posterior distributions of D_{lo}'s, we observed that several genes had a higher expression level in one organ and a lower expression in another, supporting our motivation for using the extended model (Model B) to incorporate organ specific differential expression. For example, in Table 5 the relatively large number of genes in the furthest off-diagonal positions indicate opposite (lower/higher) expression levels for those genes in different organs.

Table 5. Cross tabulation of genes according to their estimated expression groups (1-lower, 2-average, 3-higher) in the three organs (viz. K: kidney, L: liver and T: testis) under Model B.

% of genes		(T,1)	(T,2)	(T,3)	(T,1)	(T,2)	(T,3)	(T,1)	(T,2)	(T,3)	Total
(K,1)	(L,1)	0	8	691							699
	(L,2)	6	87	208							301
	(L,3)	93	50	16							159
(K,2)	(L,1)				0	25	124				149
	(L,2)				7	914	812				1733
	(L,3)				103	635	87				825
(K,3)	(L,1)							97	48	20	165
	(L,2)							119	443	111	673
	(L,3)							346	327	24	697
Total		99	145	915	110	1574	1023	562	818	155	5401

In all three organs, 90% or more of the genes with a higher than average expression also had moderate or low variance (Table 6). As expected, genes with a lower expression showed greater variation and about 70% of the genes with lower expression had low or moderate variance.

The expression measures from testis samples showed the larger share of genes within the high expression group (38.8%), compared to kidney (28.4%) and liver (31.1%). This may be due to over-representation of testis genes in the reference. Testis is known to have a greater percentage of messenger RNA than liver and kidney. When equal amounts of total RNA are used to make the reference, more of the reference ends up being from the testis since only messenger RNA is made into array probes.

Also although 5% of all genes in the testis is estimated to have high variation, only 0.3% of the 38.8% highly expressed genes were estimated to be highly varying.

Table 6. Organ-wise cross tabulation of genes (in %) according to their estimated variance and expression levels.

	Kidney data				Liver data				Testis data			
	Variance group				Variance group				Variance group			
Exp level	1	2	3	All	1	2	3	All	1	2	3	All
1	6.7	9.1	5.7	21.5	5.1	9.8	3.8	18.8	3.5	6.8	3.9	14.3
2	2.7	15.7	31.8	50.1	0.9	20.4	28.8	50.1	1.3	20.7	25.0	47.0
3	1.7	9.5	17.2	28.4	2.6	12.5	16.0	31.1	0.1	9.2	29.4	38.8
All	11.1	34.3	54.7	100	8.7	42.7	48.6	100	5.0	36.7	58.3	100

5. MOUSE MODEL (MODEL C)

Pritchard *et al.* [2001] observed that for some genes the expression varied significantly across mice. Note that both in Model A and B, by introducing an array-level effect, namely μ, a mouse-level effect has already been nested into the model. We can further extend the proposed models A and B to incorporate an additional mouse-effect. We have already established that the genes have varied expression levels across organs. Moreover, it is possible that even if a certain gene has an above average expression in a certain organ, its magnitude may vary across mice. Similarly, since the reference samples were taken from all three tissues, it is possible that the same genes might exhibit average or low expression in one or both of the remaining organs, but again its magnitude might vary across mice.

In order to account for such between-organ and between mice differences in gene expression we modeled the log-ratio of intensities of the genes in the following manner. The extended model continued to have array mean effects across all genes (i.e. μ's) as in Models A and B. Additionally each gene (I) was classified independently in each organ (O) as belonging to one of three possible expression groups (D_{IO}). Accordingly for each gene and for each mouse, an expression-level ($\theta(D_{IO}, M_J)$) with M_J indicating the mouse number for the J-th array, was added to the corresponding array mean effect (μ_{OJ}) to explain the expected log-ratio of intensities of that gene.

As in Model A, each gene was classified independently for each organ with respect to its residual variance (C_{IO}).

In summary, the conditional distribution of the log-ratio-of-intensity I_{IOJ}, given C_{IO} and D_{IO}, was assumed to be given by (with I, O, J as before),

$$I_{IOJ} = \mu_{OJ} + \theta(D_{IO}, M_J) + E_{IOJ}\text{, where } E_{IOJ} \sim N(0, 1/\tau(C_{IO})) \quad (3)$$

The prior assumptions on the original parameters were kept unchanged, with the θ's for the L-th mouse being assigned Normal priors (on appropriate ranges as before) with $\theta_{1L} < \theta_{2L} = 0 < \theta_{3L}$. The latent variables C_{IO}, and D_{IO} were assumed to be *a priori* Multinomial with corresponding probabilities (λ_C and λ_D) drawn from Dirichlet distributions. The posterior density $P(\mu, \tau, C, \lambda_C, D, \lambda_D | I)$ was then defined in a manner similar to the previous model.

5.1 Model C: Results

The estimated variance classes were comparable with the estimates obtained under Model B. The model continued to improve over the previous models, in the sense of explaining the observed variation better, by reducing the proportion of highly varying genes and increasing that of less varying genes (Table 7).

From the estimated mouse-specific expression levels it was observed that there were indeed differences in the expression levels across mice, although the differences were not significantly large.

Table 7. Cross tabulation of genes according to their estimated variance groups (1: High, 2: Moderate, 3: Low) in different organs (K: kidney, L: liver, T: testis) under Model C (in %).

% of genes		(T,1)	(T,2)	(T,3)	(T,1)	(T,2)	(T,3)	(T,1)	(T,2)	(T,3)	Total
(K,1)	(L,1)	1.2	2.7	1.3							5.2
	(L,2)	0.7	3.1	0.7							4.5
	(L,3)	0.1	0.6	0.2							0.8
(K,2)	(L,1)				0.5	0.9	0.4				1.9
	(L,2)				0.8	12.1	7.0				19.9
	(L,3)				0.3	5.3	5.3				11.0
(K,3)	(L,1)							0.4	0.4	0.6	1.4
	(L,2)							0.6	5.1	11.9	17.6
	(L,3)							0.1	6.4	31.2	37.7
Total		2.1	6.4	2.1	1.6	18.4	12.7	1.1	11.9	43.7	100.0

Similar computations as presented in Table 5 were carried out with the result that a large number of genes continued to have different expression levels across the three organs as before.

6. MODEL COMPARISON

6.1 Comparison between Models A, B and C

In most microarray data analysis normalization is done as a separate procedure without considering its effect on subsequent classification. In the three models proposed here, the data have been normalized/adjusted using increasingly available information on biological and experimental factors possibly affecting the observed data. As a result we observed gradual improvement and refinement in classification of the genes.

We have already noted that over the three models for each of the three organs the number of genes identified as having a high variance decreased, and also the share of genes with low variance increased, when moving from Model A to B and then to C.

When the behavior of any particular gene in all three organs is considered simultaneously, then collectively for all genes from Tables 2, 4 and 7 we note that the proportions of genes with high variance in all three organs was reduced across the models, with the percentages being 1.7, 1.4 and 1.2 respectively under Model A, B and C. On the other hand the percentages of genes with average or low variance in all three organs increased (A: 79.1%, B: 83.7% and C: 84.3%). This implies that employing biological and experimental hypotheses in model building better explains the observed variations in the data.

In the following we give a brief example (Figure 3) of how the models work. Of the 5401 genes, a few were selected and log-ratio-of-intensities from the kidney sample were plotted. The plot of the original data shows wide variation across arrays.

However, *a posteriori* adjusted log-ratio-of-intensities under Model A show smoothing over arrays. Recall that the model did not by itself introduce any such assumptions. Hence such smoothing implies that possibly some common factors, experimental or biological, had affected the observed log-ratio of intensities of most of the genes along with those of the selected ones. These factors, when taken care of by the array effect components in the model, smooth out the observed log-ratios.

For the selected set of genes, although the Model-A-adjusted ratios appeared much smoother than the original data, these plots were still distinctly away from zero. Clearly some of these genes have above average expression in the kidney sample, whereas others had below average. This is supported by the plots of *a posteriori* adjusted log-ratio-of-intensities under Model B. This plot shows array-wise movement towards the origin resulting in shrinkage of the plotted region.

By applying Model C some further smoothing was observed for mice 5 and 6, although very little in magnitude.

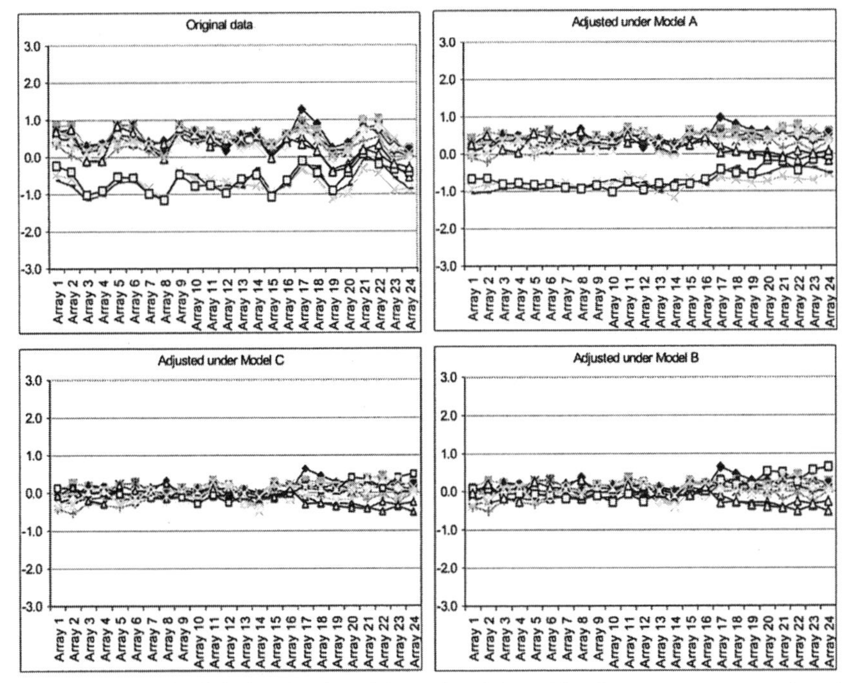

Figure 3. Plots of original and adjusted log-ratio of intensities for some genes as adjusted under Models A, B and C (Clockwise).

The proposed models gave consistent results in the sense of providing an improved explanation of observed variation with increasing complexity of the model. In principle, the models can be extended further by increasing the number of layers of the hierarchy. For practical purposes, in the absence of any specific hypothesis about essential sources of variation, amongst the proposed models, model B would be preferred. Well-established biological effects, such as tissue type, have already been accounted for in all three models that were proposed. In addition, if the biological replicates (here mice) are also suspected to cause additional variation, as suggested earlier [e.g., Pritchard et al., 2001], model C would be selected.

6.2 Comparison with Pritchard *et al.*

The genes identified by us as belonging to a large variance class were compared with those found by Pritchard *et al.* [2001]. Several such genes were identified in both studies. But we also noted that several of the genes identified by Pritchard *et al.* as significantly varying were not similarly classified by our models.

Incidentally, the genes plotted in Figure 3 are some of the genes identified as varying (in kidney) by Pritchard *et al.* [2001] using an ANOVA model whereas our analyses indicated these as having average or low variance.

This appears to be so because the emphasis in ANOVA, when performed separately on every gene, is to study the *relative* contributions of different sources of variation. In particular, Pritchard *et al.* focused on identifying genes with a relatively higher contribution of variation due to mouse, irrespective of the absolute magnitude of variation. By contrast, our main aim was to identify genes with a high degree of variation that remained unexplained by the model(s).

The genes common in the two lists generally have a high variance in the sense of magnitude, in which there is a significant contribution from between mice variation. The remaining genes from the list of Pritchard *et al.* list are probably genes which exhibit variation across mice and moderate or low residual variance. Hence future observations from these genes might vary from mouse to mouse but overall differences are expected to stay within a "tolerable" limit. The genes additionally identified here were estimated to have a high variance making future observations from them less predictable. The observed variation in these genes could not be explained any further by introducing a model for between-mouse effects.

Four liver genes and three kidney genes were analyzed by Pritchard *et al.* using real-time quantitative RT-PCR to validate expression differences between mice. Three of the 4 genes confirmed to be variable in the liver by RT-PCR were in common between the two lists. These were CisH (NM_009895), Gadd45 (NM_007836), and PRH/Hhex (NM_008245). The fourth gene confirmed in the liver, Bcl-6 (NM_09744), was not found in the large variance class of the current analysis. Instead, under both Models B and C, this gene was identified as having above average expression in liver with moderate variance. However, applying the same models to the RT-PCR data from liver assigns Bcl-6 to the large variance class.

Three kidney genes confirmed by RT-PCR as varying are CisH, Bcl-6 and complement factor D (NM_013459). Of these, CisH was identified as varying by the present analysis too. Similar to liver, our models confirmed Bcl-6 as varying in kidney tissues based on RT-PCR data but not based on microarray data. Although complement factor D was estimated to have moderate variance in kidney, its expression classification significantly changed from Model B to mouse Model C, supporting the existence of a mouse effect on this gene.

Results of Bayesian analysis of microarray data were compared for these selected genes which were identified as highly varying by Pritchard *et al*, [2001] based on microarray as well as RT-PCR data. In case of a disagreement between the two analyses, RT-PCR data was additionally used. This comparison is summarized in Table 8.

Table 8. Bayesian characterizations of selected genes which were estimated by Prichard *et al.* [2001] as highly varying in both microarray and RT-PCR data.

Organ	Gene annotation	Bayesian analysis
Kidney		
	CisH (NM_009895)	High variation
	Bcl-6 (NM_09744)	Microarray data: Avg. variation RT-PCR data: High variation
	Comp. Factor D (NM_013459)	Microarray data: Avg. variation with expression group changing from high to average from Model B to Model C indicating possible mouse effect
Liver		
	CisH (NM_009895)	High variation
	Gadd 45 (NM_007836)	High variation
	PRH/Hhex (NM_008245)	High variation
	Bcl-6 (NM_09744)	Microarray data: Avg. variation with high expression RT-PCR data: High variation

Another gene, β-2 microglobulin, which was identified by Pritchard *et al.* as varying in both kidney and in testis, was also estimated by us to have high variance in both samples under Model A. In testis under all three models this gene continued to be identified as varying. However, its estimated variance class in kidney dramatically changed when moving from Model A to Model B. Under Model B in kidney this gene was estimated to have low variance class with above average expression.

7. CONCLUDING REMARKS

Our approach to statistical modeling and analysis has been integrated in the sense that normalization and classification are being carried out simultaneously.

The normalization factors as obtained by Pritchard *et al.* were partly compared with those obtained by us under different models and were found to be comparable. Recall that our method of normalization is statistical and not deterministic as is commonly done. Additional adjustments suggested in Models B and C to incorporate above or below average expression, respectively without and with mouse specific variation, have certainly helped us to explain the observed variation. This is evident from the decrease in the number of genes estimated to have a higher variance. We emphasize the point here that all these were achieved using solely experimental and biological information and without resorting to any ad hoc or artificial adjustments.

Model A takes into account normalization for experimental factors distorting measurements of all genes on an array. Model B extends Model A by incorporating some available biological information. As an example of possible further extension Model C was proposed. With the availability of further knowledge on relevant biological factors, extended models can be formed using such information.

Classification as is proposed here helps one to reduce the dimensionality problem significantly. For example, a standard clustering algorithm would have treated the observed data from each organ as a vector of 24 dimensions. Instead we reduce the problem to two dimensions only and then perform classification with respect to the adjusted mean and residual variance of what is assumed to be 24 exchangeable observations.

It may be mentioned here that it is also possible to treat the number of groups as an unknown parameter in the model and estimated simultaneously along with the others by using techniques like reversible jump MCMC [Green, 1995; Medvedovic and Sivaganesan, 2002]. One could also consider a hierarchical model in which the variances are distributed according to some continuous distribution. In a special case this leads to Student's t-distribution for errors [Geweke, 1993]. However, even if we would consider the variances as continuous variables, in order to address the basic problem of differentiating between genes, one would finally have to resort to some sort of discretization, e.g., by introducing cut-off points.

The proposed model is based on the idea of applying a simple discrete approximation that leads to an easy implementation and intuitive interpretation of the results. In order to be able to differentiate between genes according to their variance level, suitable numbers of classes can be explored, starting from two, and depending on the required fineness/resolution of the partitions in the variance scale. Several such numbers of classes were explored here and three classes were chosen only for the purpose of illustration and easy interpretability.

Estimation of uncertainty was obtained along with classification and consequently it was unnecessary to carry out a large number of testing of hypotheses which also avoids complications of multiple comparisons.

The proposed method of analysis brought out a previously unexplored aspect of this data set. Namely the relative level of activeness of the genes as estimated by the expression groups together with their predictability, which is given by their respective variance (or precision) class estimator.

Additionally from these models, gene-level information on their respective expression levels across the organs could certainly provide useful insight into understanding the relevant regulatory networks in different organs.

REFERENCES

Baldi P, Long AD (2001) A Bayesian framework for the analysis of microarray expression data: regularized t-test and statistical inferences of gene changes. Bioinformatics 17: 509-519.

Best, N. G., Cowles, M. K. and Vines. S. K. (1995) CODA: Convergence Diagnosis and Output Analysis software for Gibbs Sampler output: Version 0.3. Cambridge: Medical Research Council Biostatistic Unit.

Dror RO, Murnick JG, Rinaldi NA (2002) A Bayesian approach to transcript estimation from gene array data: the BEAM technique. RECOMB 2002: Proceedings of the Sixth Annual International Conference on Research in Computational Molecular Biology (ACM PRESS).

Gelman A, Carlin JB, Stern HS, Rubin DB. (1995) Models for missing data. In: Bayesian data analysis. London: Chapman & Hall; pp. 439-66.

Geweke J. (1993) Bayesian treatment of the independent Student-t linear model. Journal of Applied Econometrics 8: S19-S40.

Gilks WR, Thomas A, Spiegelhalter DJ (1994) A language and program for complex Bayesian modeling. The Statistician 43: 169-178.

Green PJ. (1995) Reversible jump Markov Chain Monte Carlo computation and Bayesian model determination. Biometrika. 82: 711-732.

Ibrahim JG, Chen M-H, Gray RJ (2002) Bayesian models for gene expression with DNA microarray data. J Am Stat Assoc 97: 88-99.

Keller AD, Schummer M, Hood L, Ruzzo WL (2000) Bayesian classification of DNA array expression data, *Technical Report*, UW-CSE-2000-08-01, Dept. of Comp. Sc. & Engg., Univ. of Washington, Seattle.

Long AD, Mangalam HJ, Chan BY, Tolleri L, Hatfield GW, Baldi P (2001) Improved statistical inference from DNA microarray data using analysis of variance and a Bayesian statistical framework. J Biol Chem 276: 19937-19944.

Medvedovic M (2000) Identifying significant patterns of expression via Bayesian infinite mixture models. *CAMDA'00 : Critical Assessment of Techniques for Microarray Data Analysis*, Duke University.
Medvedovic M and Sivaganesan S. (2002) Bayesian infinite mixture model based clustering of gene expression profiles. Bioinformatics 18: 1194-1206.
Newton MA, Kendziorski CM, Richmond CS, Blattner FR, Tsui KW (2001) On differential variability of expression ratios: improving statistical inference about gene expression changes from microarray data. J Comp Biol 8: 37-52.
Parmigiani G, Garrett ES, Anbazhagan R, Gabrielson E (2002) A statistical framework for expression-based molecular classification in cancer. J Roy Stat Soc B, 64: 717-736.
Pritchard CC, Hsu L, Delrow J, Nelson PS (2001) Project normal: Defining normal variance in mouse gene expression. Proc Natl Acad Sci USA 98: 13266-13271.

SECTION VI

INVESTIGATING CROSS HYBRIDIZATION ON OLIGONUCLEOTIDE MICROARRAYS

11

QUANTIFICATION OF CROSS HYBRIDIZATION ON OLIGONUCLEOTIDE MICROARRAYS

Li Zhang, Kevin R. Coombes, Lianchun Xiao
Department of Biostatistics, The University of Texas MD Anderson Cancer Center, Houston, TX

Abstract: Cross hybridization on microarrays generates signals from unintended genes, which presents a special challenge in gene expression profiling studies since it directly leads to false positives. However, little is known about the extent of cross hybridization and why certain probes are particularly prone to cross hybridization. Recently, we developed a free-energy model of binding interactions on oligonucleotide arrays that can decompose the observed probe signals in terms of the effects of gene-specific and generic non-specific binding. We analyzed the data set provided by Affymetrix Inc., which followed a Latin square design with 14 genes spiked-in at various concentrations. Around 31 probesets show reproducible response to the spiked-in genes. In most cases, we were able to extract the amount of cross hybridization signal and identify the source, i.e., the fragments of spiked-in genes that match the cross hybridizing probes. These findings demonstrate the utility of our model for identifying spurious cross-hybridization signals and obtaining robust measure of gene expression levels.

Key words: Affymetrix, microarray, data analysis, algorithms, performance, design, reliability, and theory.

1. INTRODUCTION

On oligonucleotide microarrays commonly used in gene expression profiling studies, 25-mer DNA oligonucleotides are employed as probes to detect targeted genes with complementary sequences [Lockhart *et al.*, 1996]. However, specificity of binding on such short probes is known to be limited [Southern *et al.*, 1999]. RNA molecules with partially matched sequences are expected to cross hybridize to the probes. Current design of the microarrays uses a "perfect match vs. mismatch" approach to this problem. By design, probes with sequences perfectly complementary to fragments of targeted genes are called perfect match (PM) probes. For each PM probe, there is an accompanying mismatch (MM) probe designed to have the same sequence as the PM probe, except for the 13th base. Under the assumption that a MM signal has much less gene-specific signal, but the same amount of non-specific signal compared to a PM signal [Lockhart *et al.*, 1996], the PM minus MM value is taken to be proportional to the gene-specific signal. However, this approach is obviously flawed since the PM minus MM value is negative 30% of the times.

Recently, we developed an alternative approach based on a free energy model of molecular interactions on oligonucleotide arrays (see [Zhang *et al.*, 2003] for details). This model uses PM probes only, and assumes that there are two modes of binding on the probes: gene-specific binding (GSB) and non-specific binding (NSB).

Here, GSB refers to the formation of DNA/RNA duplexes with exact complementary sequences. NSB refers to the formation of duplexes with many mismatches between the probe and the attached RNA molecule. The number of duplexes with few mismatches should be rare because the probes are pre-selected to avoid this type of binding [Lockhart *et al.*, 1996].

The model makes two further assumptions: (1) at the completion of a hybridization experiment, a thermodynamic equilibrium state is reached between RNA molecules bound to the microarray surface and the RNA molecules in solution; and (2) binding of various RNA species are independent and noncompetitive.

The model provides a way to decompose the probe signals into GSB and NSB modes, both of which are determined by the free energies of RNA/DNA duplex formation on the microarrays, which in turn are determined by the probe's sequence. Specifically, a probe's signal is decomposed into three components using the following formulas,

$$\hat{I}_{ij} = N_j/(1+e^{E_{ij}}) + N^*/(1+e^{E_{ij}^*}) + B \tag{1}$$

where $\hat{I}_{ij}$ is the expected signal (amount of RNA bound to the probe) of the i-th probe in a probe set targeted to detect gene j; the three terms on the right represent GSB, NSB, and a uniform background, respectively. N_j is the number of expressed mRNA molecules from gene j and N^* is an unknown parameter that may be interpreted as the population of RNA molecules that contribute to NSB [Zhang *et al.*, 2003]. E_{ij} is the free energy for formation of the specific RNA/DNA duplex with the targeted gene. E_{ij}^* is the free energy for NSB, i.e., formation of duplexes with many different genes.

Given the sequence of a probe as (b_1, b_2, ... b_{25}), the values of E_{ij} and E^*_{ij} of the probe are calculated as

$$E_{ij} = \sum_{k=1}^{24} \omega_k \varepsilon(b_k, b_{k+1}) \quad (2)$$

$$E^*_{ij} = \sum_{k=1}^{24} \omega^*_k \varepsilon^*(b_k, b_{k+1}) \quad (3)$$

where ω_i and ω^*_i are weight factors that depend on the position along the probe from the 5' end to the 3' end. The $\varepsilon(b_k, b_{k+1})$ term is the same as the stacking energy used in the nearest-neighbor model [Breslauer *et al.*, 1986] to account for the interactions between adjacent bases b_k and b_{k+1}. This term also implicitly includes interactions of base pairing between the probe and an RNA molecule bound to it. $\varepsilon^*_i(b_k, b_{k+1})$ is the stacking energy for NSB. Note that Eq. (2) contains 40 unknown terms: 24 values of ω_i, and 16 values of ε_i, one for each possible combination of two bases (i.e., AA, AT, AC, AG, etc.). Similarly, Eq. (3) contains 40 unknown terms. In addition, the values of N^*, B and N_j are unknown.

Best values for all these parameters were obtained by minimizing the fitness function F to optimize the match between the expected signal intensity values ($\hat{I}_{ij}$) and the observed signal intensity values (I_{ij}) with

$$F = \frac{1}{M} \sum (\ln \hat{I}_{ij} - \ln I_{ij})^2 \quad (4)$$

where M is the total numbers of probes on an array. We have shown that the expected signals and the observed signals usually match well. Typically, the minimized value of F ranges from 0.04 to 0.11 for a data set on an array. $\hat{I}_{ij}$ and I_{ij} are well correlated. (The computer program called PerfectMatch using the model for data analysis is available online at the following URL: http://bioinformatics.mdanderson.org.)

In this study, we focused on assessing effects of cross hybridization on the arrays using our free energy model. We performed a detailed analysis of the Human data set containing 59 arrays (the array type is HG-u95a) provided by Affymetrix Inc., following a Latin square design. The samples are the same except for those 14 genes that were spiked in at concentrations ranging from 0 to 1024 pmol. The data set is ideal for studying the cross hybridization effect because there are no unknown biological variations. Any apparent changes other than the spiked-in genes must be due to cross hybridization to one of the spiked-in genes. We were interested to know how all of the probes respond to the spiked-in genes and how cross hybridization affects the probe signals.

2. RESULTS

First we identified the genes that show reproducible gene expression patterns between batches. The 59 arrays are composed of three batches, containing 20, 20, and 19 arrays (there is one array missing in the third batch), respectively. We used linear correlation of log-transformed expression levels between batches as a measure of reproducibility (Figure 1). The distribution of data points in Figure 1 seems to be symmetrical except at the top right corner. We used r12>0.9 and r23>0.9 to be the criteria for selecting the genes for further analysis. This left us with 31 probesets.

We then performed a cluster analysis of the expression profiles. As shown in Figure 2, similar gene expression profiles are grouped together. By comparing the targeted full-length gene sequences with similar expression profiles, we saw that many of them are redundantly designed to detect the same genes. For example, probeset 407_at and 37777_at both are designed to a gene called tyrosine phosphatase (accession number: X54131).

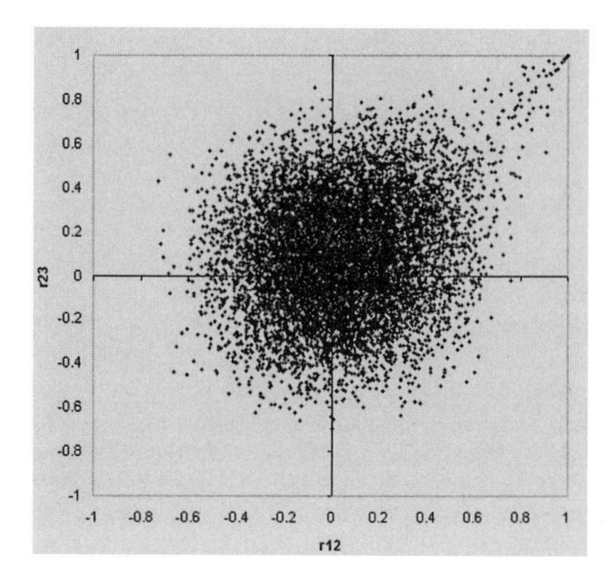

Figure 1. Correlation of expression levels between batches of experiments. Linear correlation coefficients of log-transformed expression levels between batches were plotted with each data point representing a probeset. The three batches are 1521, 1532, and 2353 series as described in sample information. r12 represents a correlation between the first and the second batch; r23 represents a correlation between the first and the second batch. The gene expression levels shown in this study were all obtained from using our model [Zhang et al., 2003]. The array type used in the experiments is HG- U95a for the human genome, which contains 12544 probesets with specified probe sequences.

The probed regions covered by the two probesets overlap by more than 90%. Three of the probe pairs in the two probesets are identical. Thus, there is no surprise that the expression of the two probesets shares nearly the same profile.

Probesets that are shown in Figure 2 but are unrelated to the spiked-in genes are supposedly the results of cross hybridization. To find out what kind of probes may produce cross hybridization signals, we performed sequence alignment between the probe sequences and the full-length sequences of the spiked-ins. We found, for example, two probes in probeset 31968_at have the following sequences:

AGCAGAACACAGTGCCTGGCATTTG and,
ATGAGCAGAACACAGTGCCTGGCAT.

The underlined fragment matches exactly to a fragment of one of the spiked-in genes (probeset name: 684_at, accession number: K02214).

We then examined the probe signals in this probeset with our free energy model. These two suspected probes are probes 5 and 6 in Figure 3, which are clearly outliers when 684_at is spiked-in at a high level (sample m at concentration of 1024 pmol); when 684_at is not spiked-in (sample q), the probe signals closely matched the expected signals from the model.

Another example was found in probeset 1032_at, which has 5 probes all containing the fragment GCAGCCGTTT. The same fragment appears twice in the full-length sequence of probeset 684_at. The 5 suspects indeed show cross hybridization in Figure 4. Thus, it seems possible to determine the amount of cross hybridization by the disparity between the model-expected signal intensity and the observed signal intensity on individual probes.

For most of the probesets shown in Figure 2, we were able to identify cross hybridizing probes similar to that shown in Figure 3 and 4; those genes are marked with a "*" in Figure 2. Probe signals in probesets 36986_at and 1552_i_at show very little change across samples. We were unable to identify the source of the minute level of cross hybridization.

We think that gene 12 in the Latin square is likely a mistake due to mislabeling. Gene 12 was reported to be 407_at according to the records. However, the probeset 37777_at also targets the same gene as mentioned earlier. We suspect that the original gene 12 may be probesets 33818_at or 33264_at, because these two probesets do not bear any identifiable sequence similarity to any of the spiked-in genes, however, they have a common gene expression profile that is expected if the Latin square design were followed exactly (Figure 5). Probesets 38502_at and 32660_at display a similar pattern of expression but changes in expression have much smaller amplitudes (Figure 2). Sources of these signals are unclear.

We also found that there may be a problem with the probe design of probeset 33264_at. Apparently, only the first 8 probes of this probeset vary among the samples, while the last 8 probes in the probeset remain non-responsive in all of the samples (Figure 6). This could be a sign of alternative splice variant since the last 8 probes correspond to a region near the last exon, which can be alternatively spliced according to a previous study [Dolnick and Black, 1996].

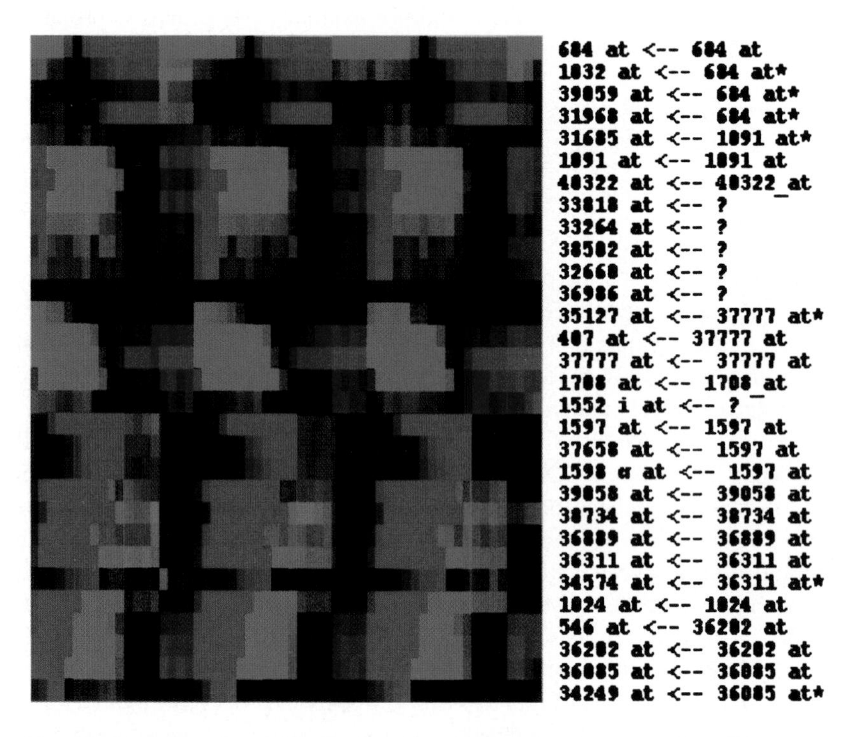

Figure 2. Clustering of expression profiles. Only those genes with reproducible patterns between batches are shown. The log-transformed expression levels are shown on a relative scale from red to green. Each row represents a probeset and each column represents a microarray. The 59 arrays are ordered by their sample names alphabetically. The periodical pattern in the figure indicates that the gene expression levels are reproducible between the batches of experiments. In each row, a probeset name is listed along with the source of the potential contributing gene identified through sequence alignments. Specific probes responding to the spiked-in genes were also determined by the disparity between the model-expected signals and the observed signals. The question marks indicate the cases in which such a source could not be identified. Changes in expression level due to cross hybridization are marked with a star symbol.

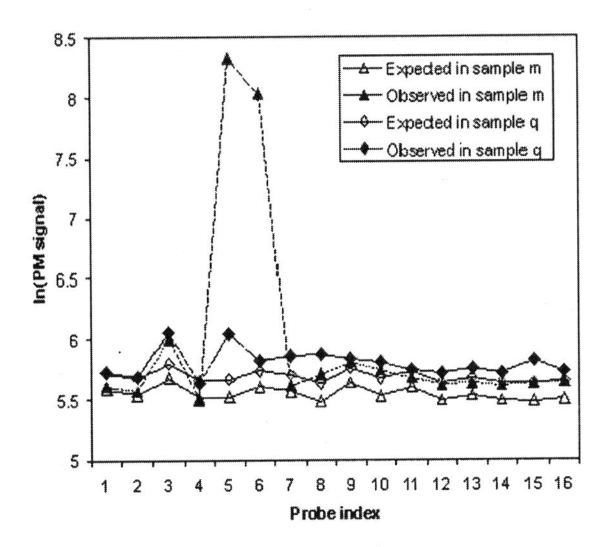

Figure 3. Spotting cross hybridizing probes in probeset 31968_at. The data were collected from batch 1521. The features presented in Figures -2 and 3 are repeatedly observed in the replicated samples but not shown.

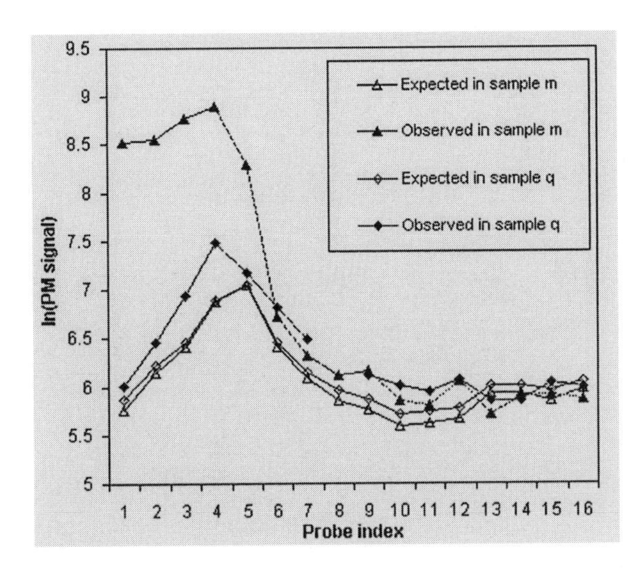

Figure 4. Spotting cross hybridizing probes in probeset 1032_at. Probes 1 to 5 contain a 10 n.t. fragment matching the sequence of 684_at. See text for details.

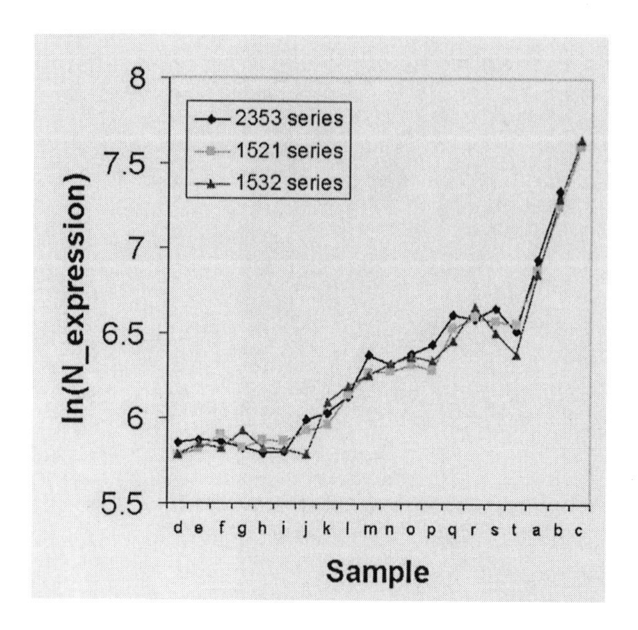

Figure 5. Expression profile measured from probeset 33264_at for all samples. Probeset 33818_at displays a similar profile, which has its minimum at sample d and maximum at sample c.

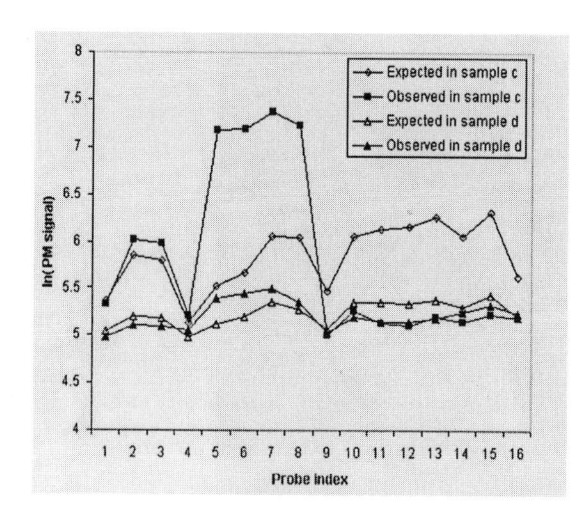

Figure 6. Probe signals in probeset 33264_at. Data were collected from samples c and d from batch 1521.

3. DISCUSSION

Our recognition of cross hybridization signals is based on binding affinities calculated from the probe sequences. The disparity between the expected signal and the observed signal allowed us to identify the individual probes that were affected by cross hybridization. It should be noted that the expected signal from the model counts both the effects of GSB and NSB. NSB in the model is an idealized concept. It assumes that there is no dominant source of cross hybridization, and it does not include any type of binding that occurs between closely matched sequences, which imposes a strong restriction on the probe design. Those probes that were found to strongly cross hybridize to the spiked-in genes can be viewed as a flaw in the probe design. We think that over-all cross hybridization signals on a microarray may be reduced by a more careful probe design avoiding sequence fragments contained in highly expressed genes.

Note that there is also an outlier removal algorithm built in the MAS 5.0, which is the commercial software developed by Affymetrix Inc. However, the outlier recognition in MAS 5.0 is based on a poorly supported assumption of normal distribution of PM minus MM signals. For instance, while MAS 5.0 is able to recognize the type of cross hybridization shown in Figure 3 because only 2 probes are outliers, it is not possible for MAS 5.0 to recognize the type of cross hybridization in Figure 4, in which 5 probes are hybridizing.

The dChip method developed by Li and Wong [Li and Wong, 2001] is also capable of detecting outlier signals on microarrays based on statistical consistency of probe signals. However, it cannot be used to detect cross hybridization discussed in this study because the cross hybridization signals are reproducible.

Cross hybridization poses a hazard in microarray measurements because it directly leads to false positives in the search of differentially expressed genes. In this study, we have provided a case study in which both the amount of cross hybridization and the source of cross hybridization can be determined. In most cases, we were able to extract the amount of cross hybridization signals and identify the source, i.e., the fragments of spiked-in genes that match the cross hybridizing probes. These findings demonstrate the utility of our model for identifying spurious cross-hybridization signals and obtaining a robust measure of gene expression levels. Note that, in a typical application of microarray experiment, the power of the method would be limited to identifying the spurious probes but not the source of cross hybridization because of the huge number of genes that can potentially contribute.

ACKNOWLEDGMENTS

Our thanks to Keith A Baggerly, Jing Wang, Ken Hess and Roberto Carta for valuable comments and suggestions on the manuscript.

REFERENCES

Breslauer KJ, Frank R, Blocker H & Marky LA. (1986) Predicting DNA duplex stability from the base sequence. Proc Natl Acad Sci U S A **83**, 3746-3750.

Dolnick BJ and Black AR (1996) Alternate splicing of the rTS gene product and its overexpression in a 5-fluorouracil-resistant cell line. Cancer Res 56(14):3207-10.

Li C & Wong WH (2001) Model-based analysis of oligonucleotide arrays: expression index computation and outlier detection. Proc Natl Acad Sci U S A 98, 31-6.

Lockhart DJ, Dong H, Byrne MC, Follettie MT, Gallo MV, Chee MS, Mittmann M, Wang C, Kobayashi M, Horton H and Brown EL (1996) Expression monitoring by hybridization to high-density oligonucleotide arrays. Nat Biotechnol 14, 1675-80.

Southern E, Mir K and Shchepinov M (1999) Molecular interactions on microarrays. Nat Genet 21, 5-9.

Zhang L, Miles MF and Aldape K (2003) A model of molecular interactions on short oligonucleotide microarrays: Implications for probe design and data analysis. Nat Biotechnol, *in press. The manuscript is also available from the authors' web site: http://odin.mdacc.tmc.edu/~zhangli.*

12

ASSESSING THE POTENTIAL EFFECT OF CROSS-HYBRIDIZATION ON OLIGONUCLEOTIDE MICROARRAYS

Seman Kachalo, Zarema Arbieva and Jie Liang

Dept. of Bioengineering and Core Genomic Facility, University of Illinois at Chicago, Chicago, IL.

Abstract: We introduce a computational method which estimates non-specific binding associated with hybridization signal intensities on the oligonucleotide-based Affymetrix GeneChip arrays. We consider a simplified linear hybridization model that should work well when the target DNA concentration is low or when the probe-target affinity is weak, and use the quadratic programming technique to estimate the parameters of this model (binding coefficients). We show that binding coefficients correlate with the degree of homology between the probe and target sequences. Detectable contribution into DNA binding was found to start from the matches of 7-8 nucleotides. The method suggested here may prove useful for the interpretation of hybridization results and for the assessment of true target concentrations in microarray experiments.

Key words: oligonucleotide microarray, cross-hybridization, linear model

1. INTRODUCTION

At the present time, DNA microarray-based comparative expression analysis has become an important tool in a variety of research areas, including cancer research, pharmacogenomics, population studies, etc. Many current microarray platforms utilize alternative probe formats bound to a solid support. This method was introduced based on the observation that single-stranded DNA binds strongly to a nitrocellulose membrane in a way that prevents strands from re-association with each other, but permits hybridization to complementary strands [Gillespie and Spiegelman, 1965]. Regardless of the probe format, all microarray based applications utilize a fundamental property of nucleic acids to re-associate separate strands in solutions in a fashion dependant on salt concentration, strand composition and sequence, as well as the degree of homology.

Hybridization of nucleic acid targets to tethered DNA probes in a multiplex or heterogeneous fashion is the central event in the detection of nucleic acids on microarrays. An immediate problem is associated with the fact that many target single strands are present in the same reaction. If their sequences are so predisposed, these target sequences can anneal with other (target and probe) strands that are not fully complementary, forming partially duplex states that are reasonably stable at assay temperatures. Obviously such "side reactions", or cross-hybridization, lower accuracy and complicate the interpretation of the microarray data. The ability to estimate the input of the cross-hybridization effect may potentially facilitate both more accurate processing of the registered hybridization intensities and more rational probe design as well.

In relation to spotted arrays, a few attempts have been made to approach the cross-hybridization issue in a more specific and quantitative manner. Riccelli *et al.* developed a new analytical method, which provides evidence of the presence of both perfectly matched and heteromorphic duplex states [Riccelli *et al.,* 2002]. The effect of the subtle sequence composition characteristics (one, two or tandem base pair mismatches and also the context surrounding the mismatch) on duplex stability and cross-hybridization propensity is under discussion. It has also been reported that for a given nucleotide probe any "non-target" transcripts >75% similar over the 50 base target may show cross-hybridization, thus contributing to the overall signal intensity. In addition, if the 50 base pair target region is marginally similar, it must not include a stretch of complementary sequence > 15 contiguous bases [Kane *et al.,* 2000].

To address the problem of cross-hybridization on Affymetrix oligonucleotide microarrays, a PM/MM approach was proposed [Affymetrix, 2002]. Each probe pair consists of a perfect match (PM) probe and a mismatch (MM) probe that is identical to the PM oligonucleotide except for a single base substitution in a central position. It is assumed that PM and MM probes are equally affected by cross-hybridization, while the PM probe has a higher level of specific hybridization. By subtracting the MM signal from the PM signal one expects to cancel the terms related to cross-hybridization and obtain a refined specific signal. However, the general assumption of equal effects of cross-hybridization on PM and MM probes is not always correct. As reported in [Naef *et al.,* 2002] about one-third of all probe pairs detect MM>PM. If the above assumption is true, these probe pairs should indicate negative gene expression.

Additional complexity is introduced by the fact that specific hybridization levels depend on the sequence of the probe. It was shown in [Li and Wong, 2001] that most individual probes are less variable between arrays than different probes within the same probe set on the same array.

This study was designed to investigate the effect of a nucleotide sequence on hybridization and the contribution of low-homologous DNA sequences into cross-hybridization.

2. DATA

We used the Human portion of the Affymetrix Latin Square dataset [Affymetrix, 2001], which can be found on Affymetrix corporate website at http://www.affymetrix.com/analysis/download or on the CAMDA website at http://www.camda.duke.edu/camda02. This dataset contains signal intensities for a total of 409,600 probes on Affymetrix HG-U95A microarray chips in 59 experiments. Experiments are divided into two groups of twenty and one group of nineteen experiments.

In each experiment fourteen labeled DNA targets with known concentrations were spiked into labeled complex targets and hybridized to the arrays. Two of fourteen targets (transcripts corresponding to the probe sets 37777_at and 407_at) are at equal concentrations in each experiment; therefore, there are only 13 distinct targets of varying concentrations in the dataset. The composition of the complex target is not specified, however, it was identical within each of the three groups of experiments. We introduced three additional variables to represent these complex targets. As the actual concentrations of complex targets had no special meaning in our study, each of these variables was assigned the value of one in one group of experiments and zero in the other two groups.

Oligonucleotide probe sequences and target definitions for HG-U95A microarray chip can be found at Affymetrix corporate website. Complete cDNA sequences for the spiked targets can be retrieved from GenBank database (http://www.ncbi.nlm.nih.gov).

3. MODELS

DNA binding to oligonucleotide probes on a microarray is a dynamic process [Tibanyenda *et al.* , 1984; Ikuta *et al.* , 1987; Wang *et al.* , 1995; Vernier *et al.* , 1996; Persson *et al.* , 1997]. The rate R_+ of DNA molecules associating with the spot is proportional to the concentration of DNA x and to the number N_{unocc} of unoccupied oligonucleotides on the microarray spot:

$$R_+ = k_+ x N_{unocc} . \tag{1}$$

The rate R_- of DNA dissociating is proportional to the amount of DNA bound to the spot or to the number N_{occ} of occupied oligonucleotides:

$$R_- = k_- N_{occ} . \tag{2}$$

Here, k_+ and k_- are the coefficients of proportionality that can depend on DNA structure, oligonucleotide sequence and many other factors. The total number of oligonucleotides per spot $N = N_{unocc} + N_{occ}$ does not change.

When equilibrium is achieved, the rates of DNA associating and dissociating become equal, i.e.:

$$N_{occ} = kxN_{unocc} , \tag{3}$$

where $k = k_+ / k_-$, or, after making all substitutions,

$$N_{occ} = \frac{kxN}{1 + kx} \tag{4}$$

Because the probe signal intensity is proportional to the amount of DNA molecules bound to the probe, the same relation can be applied for the probe signal intensity y :

$$y = \frac{kxy_{sat}}{1 + kx} , \tag{5}$$

where y_{sat} is the probe intensity in saturated state when all probe oligonucleotide molecules are bound to DNA. The dependency of signal intensity on DNA concentration is hyperbolic. However, when $kx << 1$ (i.e. when the probe signal intensity is low), it can be approximated by the linear function:

$$y = bx , \tag{6}$$

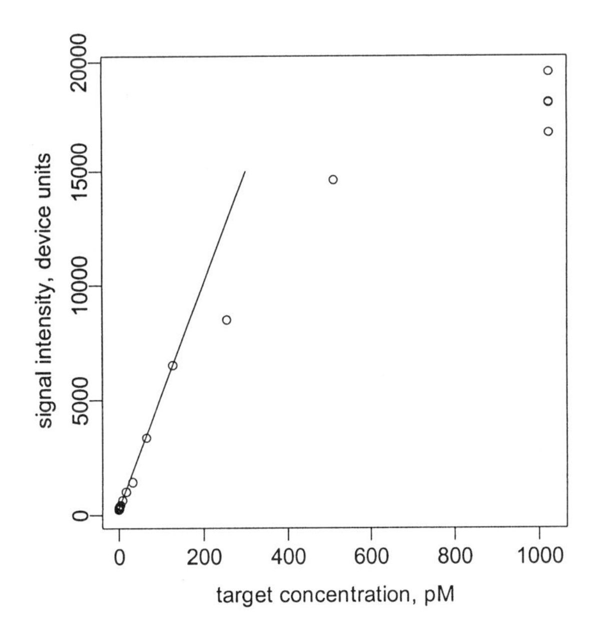

Figure 1. Dependency of probe signal intensity from target concentration. DNA transcript 684_at; probe [517:489]; first group of experiments. Probe [517:489] is specific to the transcript 684_at. The dependency can be approximated by linear function for low target concentrations.

where we define $b = ky_{sat}$ as the binding coefficient. The binding coefficient is closely related with the probe affinity effect discussed in [Li and Wong, 2001] and is equal to the logarithm of affinity effect defined in [Irizarry *et al.,* 2003].

The experimental dependency of probe signal intensity from DNA concentration is illustrated on Figure 1.

The assumption of linearity allows us to develop a linear binding model for simultaneous binding of many different DNA targets to many different probes in a series of experiments:

$$y_{ik} = \sum_j b_{ij} x_{jk} + \varepsilon_{ik} , \tag{7}$$

where $y_{ik} \geq 0$ is the signal intensity for the i-th probe in the k-th experiment, $x_{jk} \geq 0$ is molar concentration of the j-th target in the k-th experiment, $b_{ij} \geq 0$ is the binding coefficient for the j-th target and the i-th probe, and ε_{ik} is random noise.

For further comparison, we also used a random binding model that assumes that the probe signal intensities are random and independent of target molar concentrations:

$$y_{ik} = \overline{y_i} + \hat{\varepsilon}_{ik}, \tag{8}$$

where $\overline{y_i}$ is the mean signal intensity of the i-th probe in the whole set of experiments.

4. EXPERIMENTAL BINDING COEFFICIENTS

By dropping index i from (7), for each probe we can write:

$$y_k = \sum_j b_j x_{jk} + \varepsilon_k . \tag{9}$$

Provided that target concentrations x and probe signal intensities y are known for the set of experiments, binding coefficients b_j can be found as the solutions of the classical quadratic programming problem [Boot, 1964]:

minimize $\sum \varepsilon_k^{\;2}$ in (9),

subject to: $b_j \geq 0$. (10)

The program for solving the problem (10) was implemented as a combination of C++ and Matlab code. For each of 409,600 probes the program was used to calculate 16 binding coefficients (for thirteen known targets and three complex targets) from 59 data points.

The obtained binding coefficients were substituted in (9) to calculate the minimized error $\sum \varepsilon_k^{\;2}$, which was compared with the minimized error $\sum \hat{\varepsilon}_k^{\;2}$ of the random binding model (8).

As seen on Figure 2, the minimized error of the linear model is smaller than the minimized error of the random model; however, the difference is less than one order of magnitude. This can be explained by the high level of noise as well as by the nonlinearity of signal from many probes due to high probe signal intensity.

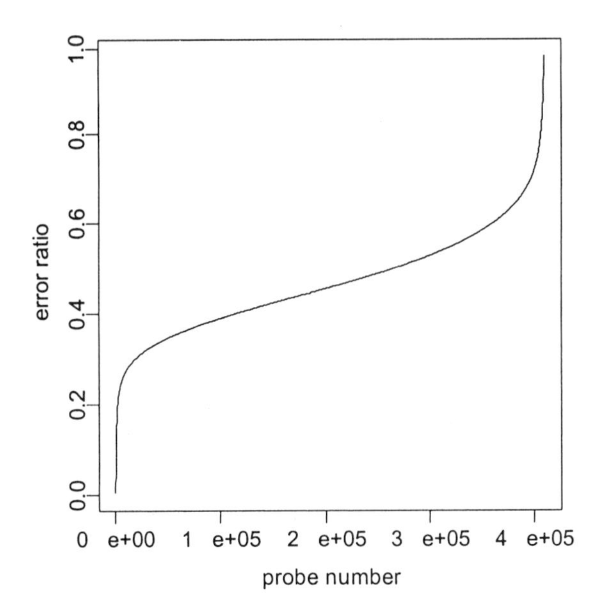

Figure 2. Sorted error ratios $\sum \varepsilon_k^2 / \sum \hat{\varepsilon}_k^2$ calculated for 409,600 probes.

For further study, a subset of 304 probes was selected for which we expected binding coefficients to be found with best accuracy. First, from the complete set, there were a few-hundred probes chosen for which the quadratic programming problem (10) solution gave the best optimization:

$$\sum \varepsilon_k^2 / \sum \hat{\varepsilon}_k^2 \leq 1/10 .$$

Next, the probes specific to, or having high similarities to the thirteen known targets were excluded from the analysis. Because of high target concentrations in the experiments, these probes were expected to demonstrate nonlinear concentration-intensity dependency.

The obtained results reveal the existence of a relationship between the binding coefficient and the degree of homology of the probe with the target nucleotide sequences. As shown on Figure 3, the correlation between the binding coefficient and the length of the longest common substring is over 60%. An almost identical relationship is observed when using the Smith-Waterman [Smith and Waterman, 1981] alignment score with various parameters instead of a common substring length.

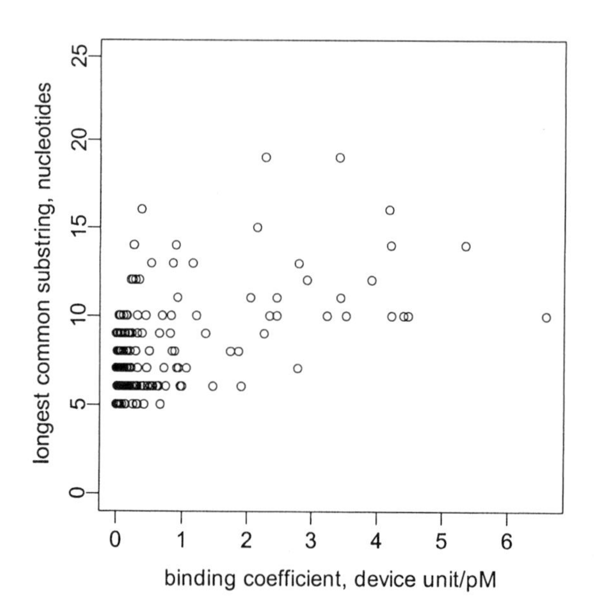

Figure 3. Binding coefficients depend on the degree of sequence similarity. Binding coefficients and longest common substring lengths for the 304 top probes and transcript 684_at are 61% correlated.

5. ESTIMATED BINDING COEFFICIENTS

As suggested by the above results, even a modest similarity may result in cross-hybridization. It is natural to think that DNA binds to the probe not only at the site of the best match, but also at the sites of weaker matches. To model this situation, many kinds of binding patterns can be introduced as multiple non-overlapping areas of similarity between the probe and target sequences that together contribute to the binding coefficient:

$$b = \sum_{a} n_a c_a + \varepsilon , \tag{11}$$

where b is the binding coefficient between any fixed probe and target, n_a - number of matches of type a found between these probe and target sequences, c_a - contribution of each pattern of type a into the binding coefficient and ε - error (not to be confused with errors in equations 8 and 9).

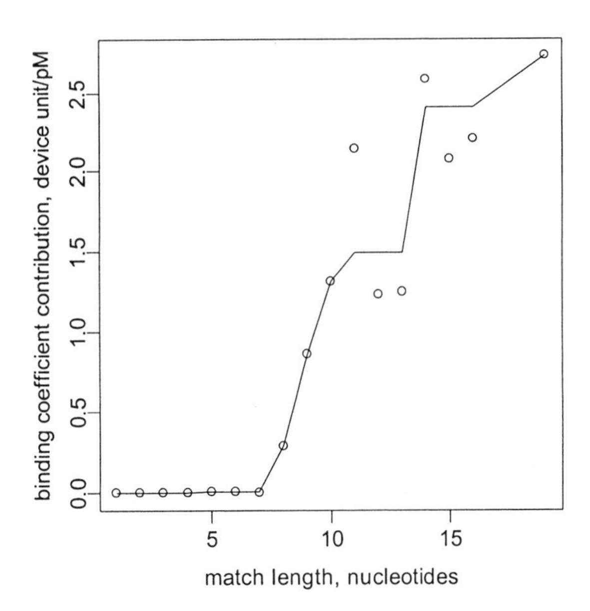

Figure 4. Contributions of perfect matches of different length into the binding coefficient, calculated for transcript 684_at and the top 304 probes. Dots are the solution of problem (13) with no additional condition; the solid line is the solution of the same problem with the additional condition (14).

Once the set of binding patterns is defined, it's easy to calculate the number of each pattern occurrence within the sequences of probe and target. If the binding coefficients are known for a number of probe-target pairs, the contribution of each binding pattern can be found by methods of quadratic programming similar to those applied for solving problem (10).

The simplest example of binding patterns can be a set of non-overlapping substrings of different lengths that are common in the probe and target sequences. Since the length of all probes on the HG-U95A microarray is 25 nucleotides, there are only 25 types of binding patterns in the set. If the binding coefficients are known for some set of probes and targets, equation (11) can now be rewritten as:

$$b_{ij} = \sum_{l} n_{ijl} c_l + \varepsilon_{ij}, \tag{12}$$

where b_{ij} is the binding coefficient for the i -th probe and the j -th target, n_{ijl} - number of matches of length l found between these probe and target sequences, c_l - contribution of each match of length l into the binding coefficient and ε_{ij} - random noise. The optimization problem to find the match contribution in this case will be:

minimize $\sum \varepsilon_{ij}^{2}$ in (12),

subject to: $c_l \geq 0$, (13)

additional condition: $c_{l+1} \geq c_l$. (14)

We used experimental values of binding coefficients for 304 probes, selected above to calculate the contributions of matches of various lengths to DNA binding. For each probe-target pair, a histogram was built for the number of non-overlapping common substrings of one to twenty five nucleotides in length. Following that, the optimization problem (13) was solved with and without additional conditions (14) using Matlab code. The problem was solved for the complete set of thirteen targets and for each target separately, revealing very similar results. Figure 4 shows the perfect match contributions obtained for one of the targets with and without additional conditions (14). A slight disagreement between these two solutions for matches longer than 10 nucleotides can be explained by the relative rarity of long matches and high level of noise, caused by that fact.

As seen from the figure, matches of length eight or greater contribute significantly to cross-hybridization. Though it's not easily apparent on the plot, contributions to cross-hybridization from matches of length seven are also detectable.

One could expect faster growth of the match contribution function with an increase in match length. Slow growth of this function for longer matches is due to the fact that probes with high similarities to targets have high signal intensities through the experiments. Because of possible non-linearity their binding coefficients may be underestimated.

Calculated match contributions were substituted back into (12) to obtain estimated binding coefficients that were then compared with experimental binding coefficients obtained in the previous section. Figure 5 illustrates the results of this comparison. The method based on the use of binding patterns performs better than the method based on match scores. We expect that this method can be further improved by using a more diverse set of binding patterns rather than the set of matches of different length. This will require, however, a larger set of experimental data.

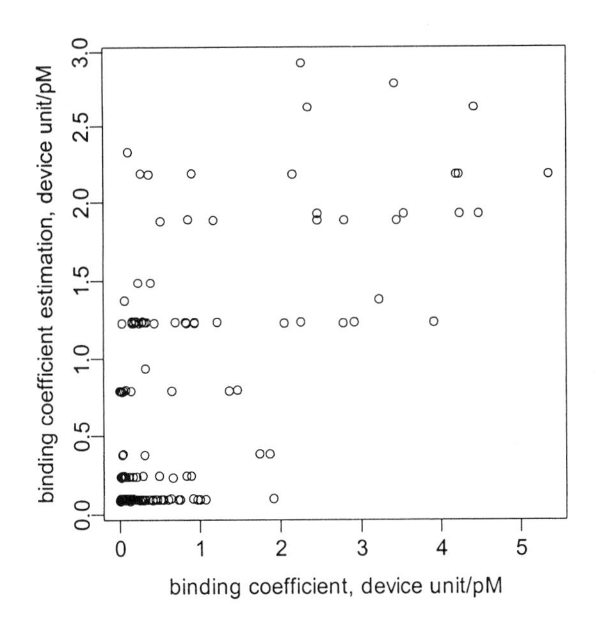

Figure 5. Experimental and estimated binding coefficients. Estimated binding coefficients for the top 304 probes and transcript 684_at are 71% correlated with experimental binding coefficients.

6. DISCUSSION

Our results demonstrate that cross-hybridization can contribute significantly to the hybridization signal, potentially introducing substantial error. By rough estimation, in the case of randomly uniformly distributed nucleotides, for any DNA transcript of 500 nucleotides in length there is about a 50% chance of a 7-nucleotide match with any 25-nucleotide probe. This suggests that any transcript, which is present in high abundance in the hybridization mixture, can affect the signal intensity for half of the probes on the microarray. In seven of nineteen possible cases, a 7-nucleotite match will cover the central nucleotide of 25-nucleotide probe. Thus, cross-hybridization differentially affects PM signal and its corresponding MM signal. This ratio is even worse for longer matches that are not as frequent as 7-nucleotide matches, but produce much stronger contribution into the signal. As reported in [Naef *et al.*, 2002], MM>PM for about one-third of all probe pairs. The only explanation of this fact is strong cross-hybridization. Though most PM/MM-based algorithms [Affymetrix, 2002;

Li and Wong, 2001] ignore such pairs, as well as other outliers, there is no guarantee that the remaining probe pairs are free from significant cross-hybridization.

A new successful algorithm not based on the PM/MM principle was recently suggested in [Irizarry *et al.*, 2003]. The model used in this algorithm can be written as

$$T(PM_{ij}) = e_i + a_j + \varepsilon_{ij}, \tag{15}$$

where T represents the transformation that corrects background, normalizes and logs the PM intensities; e_i represents the $\log_2$ scale expression value found on array i; a_j represents the log scale affinity effects for probe j; and ε_{ij} corresponds to relative error of j-th probe on the i-th array. There is an obvious tight relation between the models (15) and (7). Both models assume a linear dependency of probe signal from target concentration and imply that probe affinity to the target may be different for different probes. However, the possible effects of cross-hybridization are ignored in the model (15).

The main benefit of using the linear binding model suggested here is the opportunity to eliminate the impact of cross-hybridization. Once the binding coefficients are determined either by experiment or theoretically, finding target concentrations in (7) from the known probe signal intensities becomes a trivial linear algebra problem that can be effectively solved computationally.

The main limitation of the linear model is the fact that hybridizations should be performed at lower target concentrations than those commonly used in microarray experiments, which may result in higher relative noise level. However, the linear model (15) was shown to outperform PM/MM-based methods, probably because the concentration of spike DNA in the datasets used was about 10 times lower than in the Affymetrix Latin Square dataset.

To adopt an experiment with high target concentration, a non-linear model with more than one parameter for each probe-target pair should be applied. Its disadvantage compared to a linear model is that calculation of target concentrations from the signal intensities can be a difficult mathematical problem requiring substantially longer computational time.

ACKNOWLEDGMENTS

We would like to thank Seby Edassery and Peter Larsen from the Core Genomics Facility at UIC for their generous help during all stages of our work. We are also grateful to the members of CAMDA'02 Organizing Committee for giving us the inspiration to begin this study and for the very interesting datasets provided. This work is supported in part by a grant from The Whitaker Foundation.

REFERENCES

Affymetrix, Inc. (2001) New statistical algorithms for monitoring gene expression on GeneChip probe arrays. Technical report.

Affymetrix, Inc. (2002) Statistical algorithms description document. Technical paper.

Boot J. C. G. (1964) Quadratic Programming. North-Holland

Gillespie D., Spiegelman S. (1965) A quantitative assay for DNA-RNA hybrids with DNA immobilized on a membrane. J. Mol. Biol. 12(3):829-42

Ikuta S., Takagi K., Wallace R.B., Itakura K. (1987) Dissociation kinetics of 19 base paired oligonucleotide-DNA duplexes containing different single mismatched base pairs. Nucleic Acids Res. 26;15(2):797-811

Irizarry R., Bolstad B., Collin F., Cope L., Hobbs B., Speed T. (2003) Summaries of Affymetrix GeneChip probe level data. Nucleic Acids Res. 31:e15

Kane M., Jatkoe TA., Stumpf C., Lu J., Thomas J., Madore S. (2000) Assessment of the sensitivity and specificity of oligonucleotide (50mer) microarrays. Nucl. Acids Res. 2000 Nov 15;28(22):4552-7.

Li C., Wong W. (2001) Model-based analysis of oligonucleotide arrays: Expression index computation and outlier detection. Proc. Natl. Acad. Sci. USA, 98, 31-36

Naef F., Lim D., Patil N., Magnasco M. (2002) DNA hybridization to mismatched templates: a chip study. Phys. Rev. E Stat. Nonlin. Soft Matter Phys. 65, 040902

Persson B., Stenhag K., Nilsson P., Larsson A., Uhlen M., Nygren P. (1997) Analysis of oligonucleotide probe affinities using surface plasmon resonance: a means for mutational scanning. Anal. Biochem. 246(1):34-44

Riccelli P., Hall T., Pancoska P, Mandell K., Benight A. (2003) DNA sequence context and multiplex hybridization reactions: melting studies of heteromorphic duplex DNA complexes. J. Am. Chem. Soc. Jan 8;125(1):141-50

Smith T.F. and Waterman M.S (1981) Identification of common molecular subsequences. J. Mol. Biol. 147:195-197.

Tibanyenda N., De Bruin S.H., Haasnoot C.A., van der Marel G.A., van Boom J.H., Hilbers C.W. (1984) The effect of single base-pair mismatches on the duplex stability of d(T-A-T-T-A-A-T-A-T-C-A-A-G-T-T-G). d(C-A-A-C-T-T-G-A-T-A-T-T-A-A-T-A). Eur. J. Biochem. 15;139(1):19-27

Vernier P., Mastrippolito R., Helin C., Bendali M., Mallet J., Tricoire H. (1996) Radioimager quantification of oligonucleotide hybridization with DNA immobilized on transfer membrane: application to the identification of related sequences. Anal. Biochem. 235(1):11-9

Wang S., Friedman A.E., Kool E.T. (1995) Origins of high sequence selectivity: a stopped-flow kinetics study of DNA/RNA hybridization by duplex- and triplex-forming oligonucleotides. Biochemistry 34(30):9774-84

13

WHO ARE THOSE STRANGERS IN THE LATIN SQUARE?

Wen-Ping Hsieh[1], Tzu-Ming Chu[2], and Russ Wolfinger[2]
[1]*North Carolina State University, Raleigh, NC;* [2]*SAS Institute Inc.,Cary, NC*

Abstract: We approach the human Affymetrix Latin Square data from a classical parametric statistical modeling perspective. The first stage is to formulate a reasonable model for the probe-level data based on extant knowledge of the experimental design and technology. We present some options for this and settle on a linear mixed model for the $\log_2$ perfect match data. Upon applying this model to the data for every gene in turn, we discover that not only do the fourteen spiked-in genes appear highly significant, but that a few additional, unexpected, genes display profiles remarkably similar to those of the fourteen. Except for probe sets aimed at examining the same genes, it is likely that some short motifs might be the reason for this cross hybridization. We investigate each of these genes in more detail and offer some plausible explanations.

Key words: perfect match versus mismatch probes, mixed model, cross hybridization

1. INTRODUCTION

Microarray data analysis is a complex process involving image analysis, normalization, modeling, and clustering. Each step plays an important role in reaching accurate conclusions. To evaluate methods used for analysis, a high quality data set is very useful. The Affymetrix Latin Square data is quite appropriate for this end [Affymetrix technical report, 2002], as it provides not only a good experimental design but also the targets to make an evaluation.

In this report, we first normalize the data set and then use a linear mixed model [Chu *et al.*, 2002] to detect probe sets with significant variation between experiments, which are designed to have 14 different patterns of transcript concentration. The hybridization level of each gene is examined for each target profile. It is assumed that only the target probe sets will show the expression profile matching their respective spiked genes. Curiously, this analysis retrieves not only the putative targets that match the spiked genes, but also some unexpected probe sets that show the same profiles. We consider certain motifs to be the reason for cross hybridization and discuss some examples in detail.

2. STATISTICAL MODEL SELECTION

For these data, the experimental design is well known, although several options are available regarding which dependent variable to use in terms of the perfect match (PM) and mismatch (MM) intensity measurements. Some choices include models for paired differences [Li and Wong, 2001a, 2001b], PM-only [Chu *et al.*2002], [Lazaridis *et al.*, 2001], adjusted PM [Efron *et al.*, 2000], [Irizarray *et al.*, 2001], and both PM and MM [Lemon *et al.*, 2001], [Teng, 1999]. Some of these are on the original scale and some on a log scale, and even compromises have been recommended [Durbin *et al.*, 2002]. How to decide?

Linear reproducibility is one criterion that has bearing, and Table 1 lists the average correlation coefficient of several different intensity measurements of the fourteen spiked genes within each of fourteen experimental groups. The log transformed PM and MM values have the best consistency in this metric for most of the experimental groups. Based on this evidence, and the fact that the raw intensities represent pixel counts ranging heterogeneously over several orders of magnitude, a log transformation is justifiable. For our modeling efforts, we prefer log base 2 so that a unit difference on this scale can be interpreted as a two-fold change in the original scale. Furthermore, if the amount of cross-hybridization for an individual probe is proportional to the observed signal, and the constant of proportionality remains stable across the experiment, then this constant will cancel out for any differences taken on the log scale.

Figure 1 plots replicate values of $\log_2$(PM) and $\log_2$(MM) against one another for the three chips in experiment A. This plot reveals some potential data quality issues that should be addressed. In particular, outliers that appear well away from the main diagonal represent inconsistent measurements that should be handled carefully and potentially filtered out of the analysis.

Regarding how to handle the mismatch data, Figure 2, from a randomly chosen chip (1532e99hpp_av04), shows how strongly $\log_2$(MM) is correlated with $\log_2$(PM). MM is clearly picking up true signal and is subject to noise, and therefore subtracting it directly from PM is likely not the most optimal way to proceed.

Table 3. Average correlation coefficient across replicates within the experiment groups

	Average Correlation Coefficient				
Experiment	Log PM	Log MM	PM	MM	PM-MM
A	.9811	.9876	.9712	.9921	.9213
B	.9940	.9911	.9918	.9905	.9798
C	.9926	.9917	.9888	.9843	.9887
D	.9934	.9918	.9880	.9845	.9823
E	.9944	.9915	.9882	.9878	.9881
F	.9905	.9860	.9776	.9823	.9707
G	.9928	.9902	.9883	.9878	.9861
H	.9934	.9890	.9901	.9825	.9865
I	.9904	.9830	.9864	.9804	.9793
J	.9954	.9939	.9912	.9891	.9874
K	.9938	.9770	.9871	.9788	.9785
L	.9948	.9931	.9215	.9868	.8644
M,N,O,P	.9957	.9912	.9898	.9855	.9862
Q,R,S,T	.9952	.9896	.9903	.9883	.9819
Average	.9927	.9891	.9822	.9858	.9701

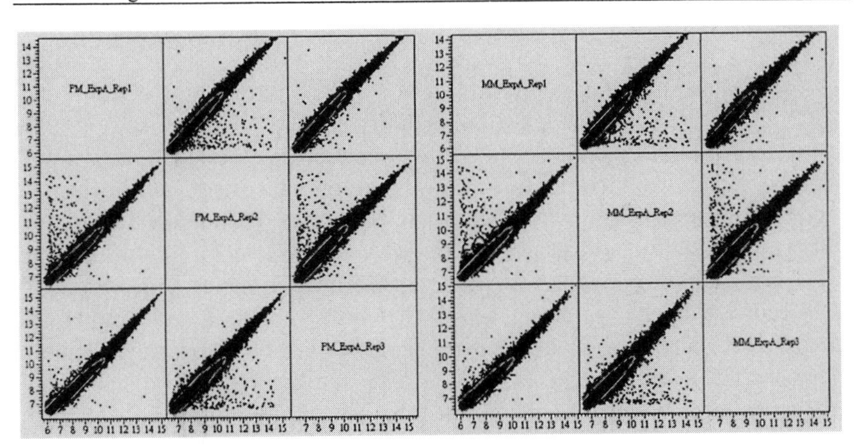

Figure 1. Scatter plot matrix of $\log_2$ PM (left) and $\log_2$ MM (right) for the three replicate chips in experiment A.

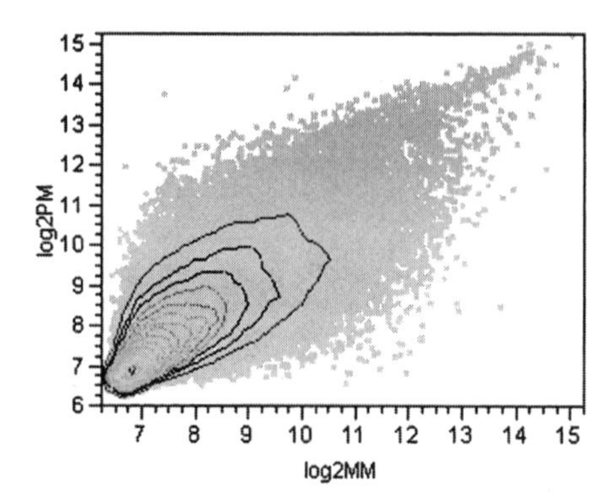

Figure 2. Plot of $\log_2$(MM) versus $\log_2$(PM) for chip 1532e99hpp_av04 in Experiment E. Each curve represents a contour of the bivariate density for $\log_2$(MM) and $\log_2$(PM).

Based upon the preceding considerations, we employ $\log_2$(PM) as a dependent variable in a linear modeling context. Research on methods for including MM as an additional dependent variable in a bivariate fashion is underway and will be reported elsewhere. Our analysis follows methods reported in [Wolfinger *et al.*, 2001] and [Chu *et al.*, 2002]. In particular, we employ the following two models in turn:

$$\log_2(PM_{ijkg}) = E_i + A_{ij} + \varepsilon_{ijkg} \qquad (1)$$

$$R_{ijkg} = E_{ig} + P_{kg} + A_{ijg} + \varepsilon_{ijkg} \qquad (2)$$

In Model (1), the symbols *PM, E, A,* and ε represent perfect match probe intensity, global experiment effect, global chip random effect, and stochastic error term, respectively. Here, "global" emphasizes that the corresponding effect applies across all genes. To be more precise, PM_{ijkg} means the intensity of the *j*th replicate of the *i*th experiment for the *k*th perfect match probe of the gth gene. In Model (2), the symbols *R, E, P, A,* and ε represent the residual calculated from Model (1), gene-specific experiment effect, gene-specific probe effect, gene-specific chip random effect, and stochastic error term, respectively. Model (1) is fitted once to jointly normalize all of the data, and model (2) is fitted separately for each gene.

3. RESULTS

We fit the preceding models to the data for all 12,626 genes using SAS Proc Mixed. After this modeling, we filtered outliers for standardized residuals greater than a certain threshold and then collected various output statistics. The fourteen spiked genes each had wildly significant results, with overall $-\log_{10}$(p-values) around 300. Figure 3 displays the profiles for these genes on a standardized scale. The resulting subtle S-shape matches that seen in other analyses.

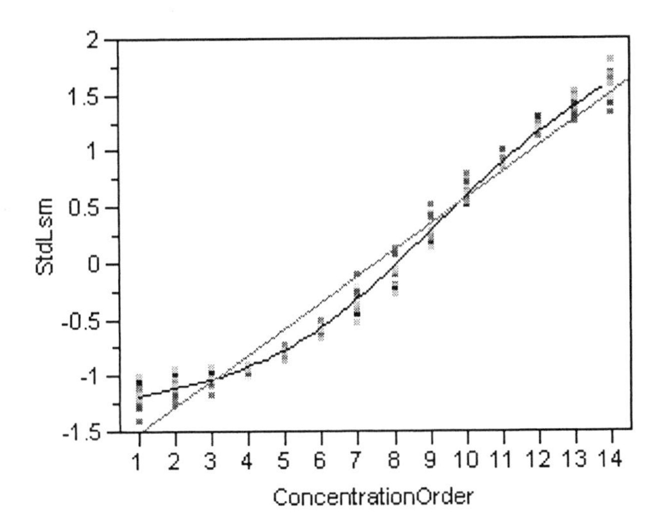

Figure 3. Overlay of standardized least squares mean profiles for the 14 spiked genes. The straight line and the curve represent linear and smooth nonparametric fitted lines respectively.

What surprised us, however, was the significance of several genes in addition to the putative fourteen. These are listed in Table 2. Expression profiles of 14 target genes are shown in Figure 4 and the top five unexpected genes including the one that was claimed to be missing are displayed in Figure 5. A few have obvious explanations, but others do not. One example of the latter is 1032_at, which has an expression profile matching that of the spiked gene 684_at.

Table 2. Unexpected significant genes from the mixed model analysis

Probe Set	Target	Remark
33818_at	AC004472	should be in Latin Square as Transcript #12
546_at	S76965	same profile as 36202_at
1598_g_at	L13720	same profile as 1597_at
37658_at	L13720	same profile as 1597_at
1032_at	U11872	same profile as 684_at

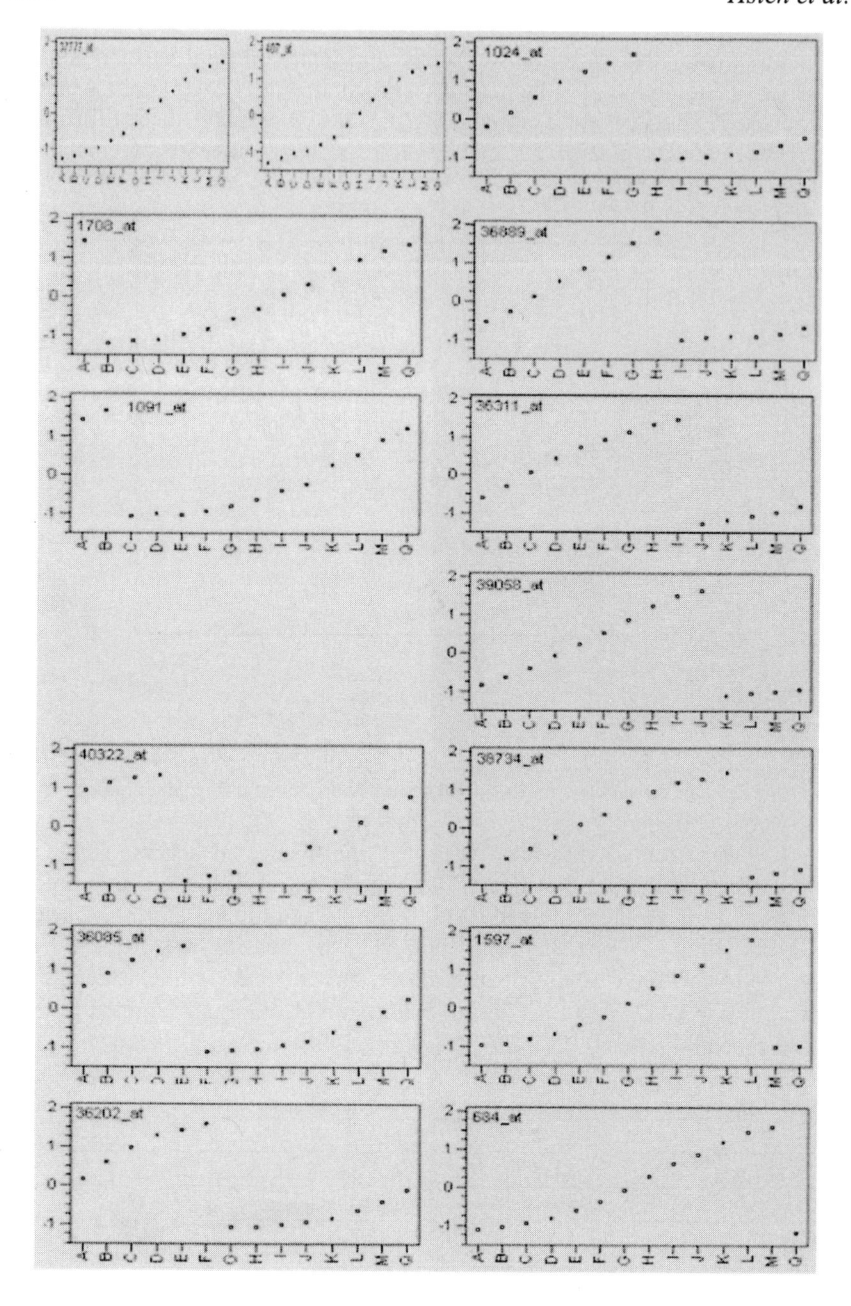

Figure 4. Standardized least squares mean profiles of significant genes from the mixed-model analysis.

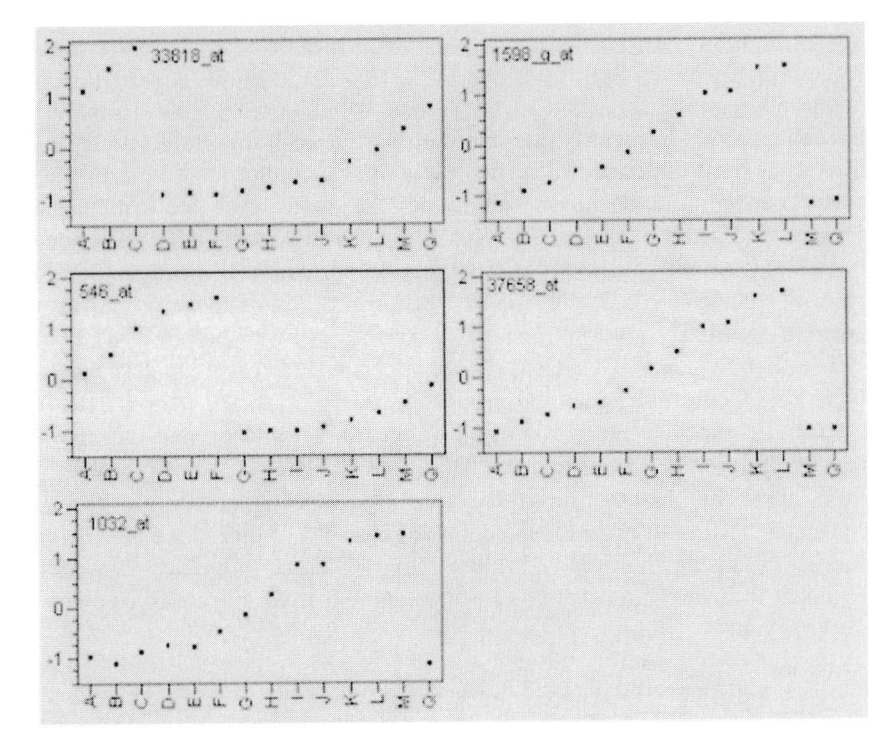

Figure 5. Standardized least squares mean profiles of unexpected genes from the mixed-model analysis.

We discovered the unexpected genes in Table 2 by statistically screening the expression profiles of the target transcripts with that of all probes in the U95A chip. Empirically, the intensity measurement is not quite linearly proportional to the transcript concentration level at high and low concentrations. To reflect the observed intensity measurement patterns, we used least square means of the experiment effects in model (2) as queries to retrieve probes with similar expression profiles. We first take averages of expression levels across replicates for each experiment and then calculate the correlation coefficients of those 14 average values with the 14 least square mean values. The higher the correlation coefficient, the higher the similarity between our target gene and the sequence to which it cross hybridized. Once obtained, the selected matches are obvious by visual comparison of Figures 4 and 5.

A brief investigation of the five genes in Table 2 produced simple explanations for their observed expression profiles, except for that of probe set 1032_at. We discovered that five probes of 1032_at have correlation coefficients higher than 0.99 with profile of 684_at while the other 11 probes have correlation coefficients less than 0.5. Since the five highly correlated probe sequences have significant overlap with each other, we expected part of this overlapping sequence to be definitive, and used Gibbs Sampling [Lawrence, 1993] to identify possible motifs. For each threshold setting on correlation coefficients, we identified the longest common motifs of all the probes chosen and manually identified the most probable candidate sequences. Not all of the probes have similar sequences as those of probe set 1032_at, but we found two of them that supported our conjecture. They are displayed in Table 3, and the expression profiles of those probes are shown in Figure 6.

The first sequence, GCAGCCGTTT, appears in seven probe sequences of the U95A chip other than those in probe set 1032_at. Only three out of seven match the sequence similarity profile of target 684_at, however they are not so strong as the second motif CCGTTTCTCCTTGGT in probe set 39059_at. This 15-base motif has a similar counterpart in the probe sequences of 1032_at but with an additional base T. If this single base T is allowed to mismatch when hybridizing to the target sequence, then the alignment in Figure 7 seems to be a promising reason for the observed cross hybridization.

Table 3. Probe sequences with expression profile matching that of target 684_at from netaffy.com. Highlighted motifs can be aligned to the sequence in gene K02215 (target of 684_at).

Probe Set	Position	Probe	Probe Sequence	Correlation coefficient of expression profile with the target profile 684_at
1032_at	46	1	agaatatGCAGCCGTTTtctccttc	0.997
1032_at	48	2	aatatGCAGCCGTTTtctccttcct	0.997
1032_at	49	3	atatGCAGCCGTTTtctccttcctg	0.998
1032_at	51	4	atGCAGCCGTTTtctccttcctggg	0.994
1032_at	52	5	tGCAGCCGTTTtctccttcctgggt	0.994
34404_at	1600	11	tgGCAGCCGTTTcttaacatgttga	0.875
38729_at	1878	5	agactcctggGCAGCCGTTTtcctc	0.888
38729_at	1885	6	tggGCAGCCGTTTtcctcatccttt	0.824
1402_at	2105		gagtGCAGCCGTTTcagaagaaaac	< 0.5
32616_at	2229		acatctgagtGCAGCCGTTTgagaa	< 0.5
32616_at	2238		tGCAGCCGTTTgagaagaaaacatc	< 0.5
40261_at	1338		gacatGCAGCCGTTTcggggtagat	< 0.5
39059_at	2305	3	gtgcgCCGTTTCTCCTTGGTagcgt	0.996
39059_at	2310	4	CCGTTTCTCCTTGGTagcgtgcacg	0.991

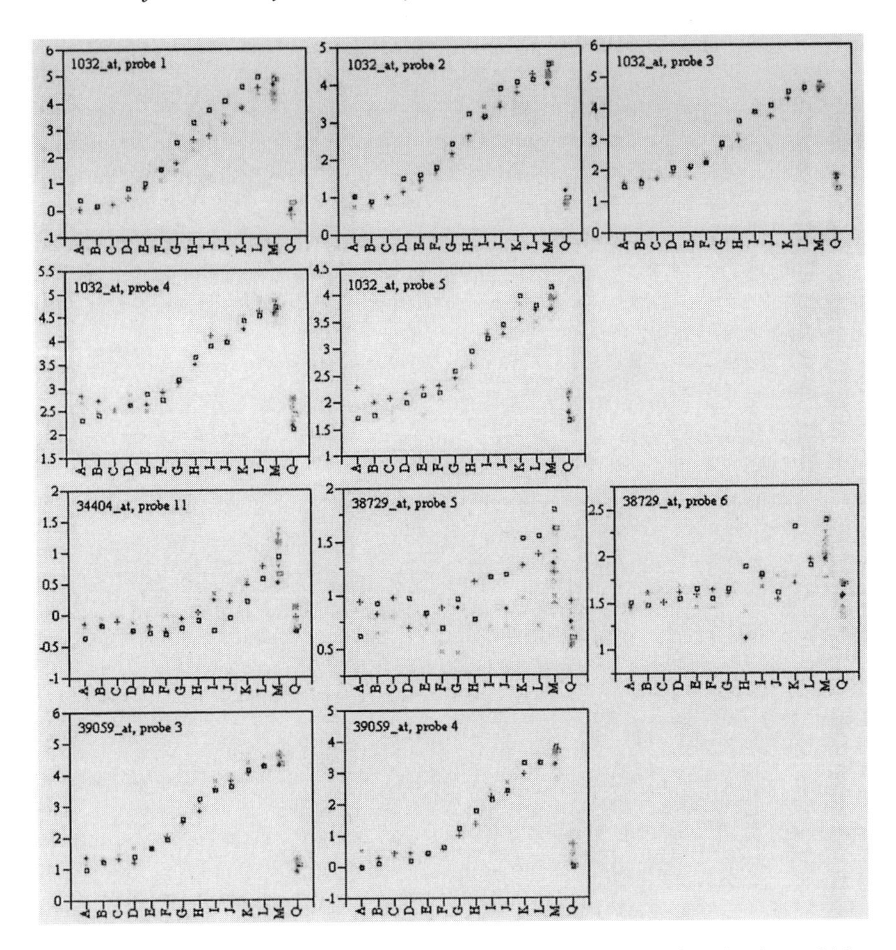

Figure 6. Expression profiles of probes listed in Table 3. Each plot contains 59 points, which represent 59 chips and there are two to twelve replicates for each experiment.

```
1032_at_1       ---------------------------ATGCAGCCGTTTTCTCCTTCCTGGG--------
1032_at_2       ----------------------------TGCAGCCGTTTTCTCCTTCCTGGGT-------
1032_at_3       ----------------------AGAATATGCAGCCGTTTTCTCCTTC-------------
1032_at_4       ------------------------AATATGCAGCCGTTTTCTCCTTCCT-----------
1032_at_5       -------------------------ATATGCAGCCGTTTTCTCCTTCCTG----------
39059_at_3      ----------------------------GTGCGCCGTTT-CTCCTTGGT---AGCGT---
39059_at_4      ---------------------------------CCGTTT-CTCCTTGGT---AGCGTGCA
K02215          CTTCTAATGAGTCGACTTTGAGCTGGAAAGCAGCCGTTT-CTCCTTGGTCTAAGTGTGCT
                                                 ****** ******
```

Figure 7. Sequence alignment for probes of 1032_at and 39059_at with K02215 (target of 684_at).

4. CONCLUSION

A linear mixed model of $\log_2$(PM) is a powerful method for assessing significance of gene expression profiles. For the Affymetrix Latin Square data, it detected all fourteen spiked genes with extremely high precision, as well as five additional "strangers". One of the five, 1032_at, did not have an initially obvious explanation, but after a more detailed motif finding exercise, we were able to find a few motifs that likely caused the cross hybridization. Additional spiked-in experiments like this one will be useful for further insights into probe performance and design.

REFERENCES

Affymetrix, (2001), New statistical algorithm for monitoring gene expression on GeneChip probe arrays. http://www.affymetrix.com/support/technical/technotes/statistical_ algorithms_technote.pdf

Chu, T., Weir, B., and Wolfinger, R. (2002). A systematic statistical linear modeling approach to oligonucleotide array experiments. Math. Biosci. 176, 35-51.

Efron, B., Tibshirani, R., Goss, V., and Chu, G. (2000). Microarrays and their use in a comparative experiment, Technical Report, Stanford University.

Durbin, B., Hardin, J., Hawkins, D., and Rocke, D. A variance-stabilizing transformation for gene expression microarray data. ISMB 2002. in press.

Irizarry, R. A., Hobbs, B., Collin, F., Beazer-Barclay, Y., Antonellis, K., Scherf, U. and Speed, T. (2001).Exploration, normalization, and summaries of high density oligonucleotide array probe level data.

Lawrence CE, Altschul SF, Boguski MS, Liu JS, Neuwald AF, Wootton JC. (1993), Detecting subtle sequence signals: A Gibbs sampling strategy for multiple alignment. Science 262:208-214.

Lazaridis, E. N., Sinibaldi, D., Bloom, G., Mane, S., and Jove, R. (2001). A simple method to improve probe set estimates from oligonucleotide arrays. Math. Biosci. 176, 53-58.

Lemon, W. J., Palatini, J. J. T., Krahe,R., and Wright , F. A. (2001). Theoretical and experimental comparisons of gene expression indexes for oligonucleotide arrays. (in press)

Li, C. and Wong, W. H. (2001a). Model-based analysis of oligonucleotide arrays: Expression index computation and outlier detection. Proc. Nat. Acad. Sci. USA 98(1), 31-36.

Li, C. and Wong, W. H. (2001b). Model-based analysis of oligonucleotide arrays: Model validation, design issues and standard error application. Genome Biol 2(8), research0032.1-0032.11.

Teng, Chi-Hse, Nestorowicz, A., and Reifel-Miller A. (1999). Experimental designs using Affymetrix GeneChips. Nature Genetics 23, 78, DOI: 10.1038/14415 Poster Abstracts

Wolfinger, R. D., Gibson, G., Wolfinger, E. D., Bennett, L., Hamadeh, H., Bushel, P., Afshari, C., and Paules, R. S. (2001). Assessing gene significance from cDNA microarray expression data via mixed model. J. Comp. Bio., 8, 625-637.

SECTION VII

FINDING PATTERNS AND SEEKING BIOLOGICAL EXPLANATIONS

14

BAYESIAN DECOMPOSITION CLASSIFICATION OF THE PROJECT NORMAL DATA SET

T. D. Moloshok[1], D. Datta[1], A. V. Kossenkov[1,2], M. F. Ochs[1,*]

[1]*Division of Basic Science, Fox Chase Cancer Center, Philadelphia, PA* and [2]*Moscow Physical Engineering Institute, Moscow, Russian Federation*

Abstract: The process of development and maintenance of tissues is complex. Key genes have been identified that serve as master switches during development, regulating numerous other genes and guiding differentiation and the development of complex structures. Within cells comprising different tissue types, the genomic complements of genes remain the same. Understanding what differentiates one tissue from another requires a different level of analysis, one performed at a functional rather than structural level. We have used a new version of the Bayesian Decomposition algorithm to identify tissue specific expression. The expression patterns which result were analyzed in terms of the ontological information on the processes in which a gene is involved by exploration of the Gene Ontology database using automated software for data retrieval and analysis.

Key words: Bayesian methods, gene expression, gene ontology

1. INTRODUCTION

Evolution has led to the rise of multicellular organisms comprised of multiple cell types with specialized roles. Each cell develops from a progenitor and all cells derive from a single fertilized egg. The replication of the cellular DNA is strongly constrained, so that each cell contains essentially exact copies of the full genetic complement. These genes are regulated in a tissue specific manner however, through various processes such as DNA methylation and the control of transcription by DNA binding and translation complexes. As a result, the differences between tissues are to

* author to whom correspondence should be addressed

be found not within the sequence of the DNA, but instead in the expression of specific genes and the translation of the expressed mRNAs into proteins.

The differences in gene expression between multiple tissue types can be utilized in a number of ways. The most common method is to look for a "fingerprint" that identifies a type of tissue from the background of multiple other types, often different types of cancer or malignant tissue against normal tissue from the same original type. Early work in this area applied neighborhood analysis [Golub *et al.*, 1999], clustering [Alizadeh *et al.*, 2000], support vector machines [Brown *et al.*, 2000], multidimensional scaling [Bittner *et al.*, 2000], artificial neural networks [Khan *et al.*, 2001], and recently partial least squares with the inclusion of survival data [Park *et al.*, 2002]. Most of these methods mix a discovery step to identify patterns and a classification step to use those patterns to separate tissue types.

Another type of analysis has the primary aim of recovering underlying biological information from the data. The goal of these methods is discovery of novel behavior. When the goal of the analysis is identification of specific mRNA species that are differentially regulated, methods in use include statistical methods such as ANOVA analysis [Kerr *et al.*, 2001; Kerr *et al.*, 2002], maximum likelihood estimation [Ideker *et al.*, 2000], as well as standard mean and standard deviation methods. When the goal is to identify coregulation, methods look for correlations in expression and include clustering [Eisen *et al.*, 1998; Heyer *et al.*, 1999; Getz *et al.*, 2000; Kerr *et al.*, 2001; Hanisch *et al.*, 2002], principal component analysis [Alter *et al.*, 2000], and self-organizing maps [Tamayo *et al.*, 1999].

In actual biological systems, the behavior is more complex than most analysis methods can take into account. While identification of a single gene which is differentially regulated between two conditions is well-defined, the grouping of genes into coregulation groups is complex, since many genes are multiply regulated and therefore naturally belong to multiple coregulation groups. The ability to assign genes to multiple "clusters" and identify multiple regulation was the driving force behind the application of the Bayesian Decomposition algorithm to microarray data [Bidaut *et al.*, 2002; Moloshok *et al.*, 2002]. We have extended Bayesian Decomposition to allow classification to play a role, so that the additional information on tissue type, such as in the Project Normal data [Pritchard *et al.*, 2001], can be utilized.

2. METHODS

Tissue development is a complex process involving the differential regulation of the genetic complement of a cell to produce specialized cells devoted to tasks necessary to the purpose and environment. Cells within organs and specialized tissues have very different functions in the mouse. Nevertheless, the cells within these tissues naturally also share considerable functions involving metabolism, creation of protein substrates, maintenance of chromosomal integrity, and numerous other "routine" processes. As such, the expression patterns within each tissue will contain both a unique set of genes necessary to the specialized purpose of the organ and additional sets of genes involved in the normal operations common to all eukaryotic cells in multicellular organisms. Through evolutionary development, the genes which provide part of a unique function in an organ may also be "borrowed" to provide a different function in different cells, including the possibility of developing a ubiquitous function necessary in all cells.

In the Project Normal data set [Pritchard *et al.*, 2001], the situation is analogous. Here three specific tissue types are represented within the data (kidney, liver, and testis). In each tissue, some genes are likely to be more strongly upregulated than in other tissues, with the extreme situation being genes that are expressed only within the specific tissue type (e.g. genes utilized uniquely in spermatogenesis are likely to be expressed only in the testis). Again there will exist a strong background level of transcription of genes involved in routine metabolic and cell maintenance processes, with this background likely varying between tissue types non-stochastically and between individual mice stochastically. The need, then, is to identify tissue specific expression behavior while not interpreting routine expression unrelated to tissue specific transcription as tissue related. We have used a modified version of Bayesian Decomposition to resolve this problem.

Bayesian Decomposition is a matrix decomposition algorithm that allows the encoding of additional prior information within a Bayesian framework [Sivia, 1996]. The input is a set of data in the form of a matrix, **D**, which describes the measurements of expression levels for genes (rows) over various conditions (columns). In addition, there is a matrix $\boldsymbol{\varepsilon}$ which provides estimates of the uncertainty or noise for each individual measurement in **D**. From this data, two matrices are constructed such that

$$\mathbf{D} = \mathbf{AP} + \boldsymbol{\varepsilon} \qquad [1]$$

where **P** contains k rows giving k patterns within the data across the conditions and **A** provides a measure of how strongly each gene contains each pattern. Presently k must be estimated outside the application of the algorithm, although the results of the analysis can guide the ultimate choice of k as demonstrated below.

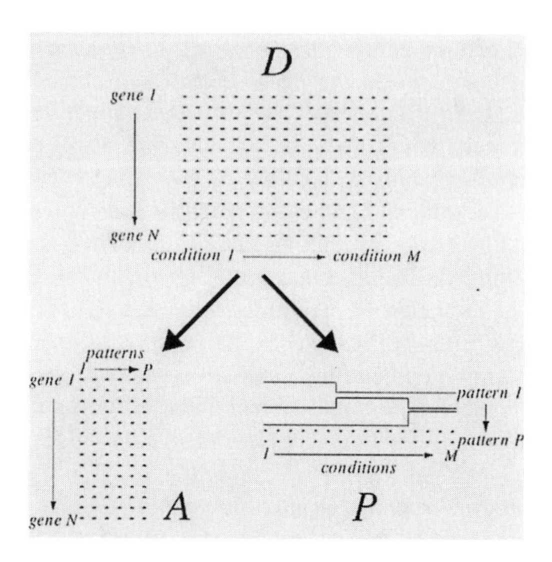

Figure 1. The Matrix Decomposition performed by Bayesian Decomposition when applied to a classification problem. The matrices **A** and **P** are identified simultaneously during a Markov chain sampling using Bayesian statistics. In this figure, three patterns are shown related to class, with the first two patterns having six members each and the third pattern having four members. The additional patterns are unconstrained. All patterns are normalized such that each row totals 1 when statistics are generated.

Because Equation 1 is mathematically degenerate, multiple solutions for **A** and **P** could exist. However, if there are more elements in **D** than in **A** and **P**, and if there is a solid underlying mathematical model of the biological system encoded, we have found that usually only one solution exists. In order to identify this solution, the algorithm encodes a Markov chain Monte Carlo method with a Gibbs sampler [Besag *et al.*, 1995] using Bayesian estimates of the probability [Sibisi *et al.*, 1997]. The possibility of being trapped in a false maximum in the probability distribution is reduced by performing simulated annealing during equilibration [Geman *et al.*, 1984]. Details of the application of the algorithm with no correlations enforced between the elements in **A** or in **P** have been published previously for analysis of the yeast cell cycle [Moloshok *et al.*, 2002] based on data from Stanford [Cho *et al.*, 1998; Spellman *et al.*, 1998] and for analysis of yeast deletion mutants [Bidaut *et al.*, 2002] based on data from the Rosetta compendium [Hughes *et al.*, 2000].

In order to make use of the additional information provided in the Project Normal data set (i.e. the knowledge of the tissue source of the mRNA), the Bayesian Decomposition algorithm has been modified to allow correlations across the rows of the **P** matrix. Effectively, for each class identified in the data, a pattern is reserved which is of value *1/N* when the condition is a member of the class and 0 when it is not a member of the class, with *N* being the number of members of the class. This creates a normalized step function which is nonnegative only if the condition is a member of the class (see Figure 1). For this data set, there are three classes as reflected in the figure, with the first six elements in class 1 (kidney), the second six elements in class 2 (liver), and the final four elements in class 3 (testis). The height of the step function for class 3 is larger reflecting the lower number of members and the normalization condition noted above. In addition to these "class" patterns, there are additional unconstrained patterns that allow the algorithm to identify correlations between the conditions and genes that do not relate to the classes of conditions (i.e., a background of expressed metabolic genes present in all samples).

The additional patterns are the key to the application of Bayesian Decomposition to the classification problem. As with previous applications of Bayesian Decomposition, the goal is to handle situations where genes are multiply assigned. In this case, genes could be assigned not only to multiple classes (e.g., a protein kinase that functions in liver and kidney), but also to background processes that are not in any way linked to individual classes, such as routine metabolism that varies between different samples due to local conditions (e.g., between different samples from mouse kidney).

The revised Project Normal data were downloaded from the CAMDA web site (http://www.camda.duke.edu/CAMDA02/). The raw data consisted of gene expression levels measured in quadruplicate for all samples [Pritchard *et al.,* 2001]. The expression levels of control and experimental measurements were determined by subtracting the median background intensity from the median foreground intensity, and these values were used to determine expression ratios for each sample. The mean expression ratio was then calculated by averaging the ratios from all the reliable measurements in the replicates. The standard deviation of the mean across these ratios was used as the uncertainty in the measurements. For data points for which one or fewer ratios were available or for data where the average ratio was negative, the ratio was set to 1.0 and the uncertainty to 100, effectively ensuring that the data point would not influence the model. In addition, the third and fourth testis data were removed as their average ratios across all genes were 2.74 and 3.11 respectively, as opposed to an average of 1.02 for all other conditions with these data removed. The final data set contained 97,979 points with four replicates, 1013 with three replicates, and

146 with two replicates. Only 24 data points were eliminated due to lack of successful replication (not counting the third and fourth mouse testis data).

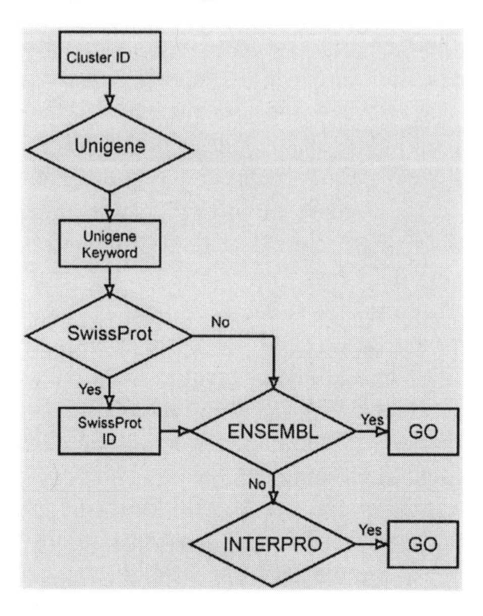

Figure 2. The annotation pipeline used to determine the Gene Ontology information. For each clone spotted on the array, updated Unigene information was retrieved. The Unigene Keyword was used to search the Swiss-Prot database. If a match was found the Swiss-Prot ID was used to retrieve gene ontology information for the clone from Ensembl or Interpro databases. If no Swiss-Prot ID was identified, the Unigene Keyword was used instead.

Two data sets were produced. The first data set utilized only data (excluding controls) for which Gene Ontology information was available [Ashburner *et al.,* 2000]. The gene ontology information was retrieved using the automated sequence annotation pipeline [Kossenkov *et al.,* 2003], as depicted in Figure 2. The full Gene Ontology hierarchy was retrieved for each spotted clone in this way and all levels of the hierarchy were used in the analysis. The final edited data set comprised 827 genes (out of 5305 with clone IDs) and 16 conditions (6 kidney, 6 liver, and 4 testis). The second data set comprised an unedited set of data (results not discussed here).

Bayesian Decomposition was used to analyze the data, positing 4, 5, and 6 patterns in the data. The results were interpreted in terms of fit to the data and gene ontology assignment of genes within each pattern. The goal of the analysis was to identify the functional processes that are enhanced in each tissue, as well as to identify other processes shared between the tissues.

These other processes are expected to show stochastic variation due to uncontrollable biological variation.

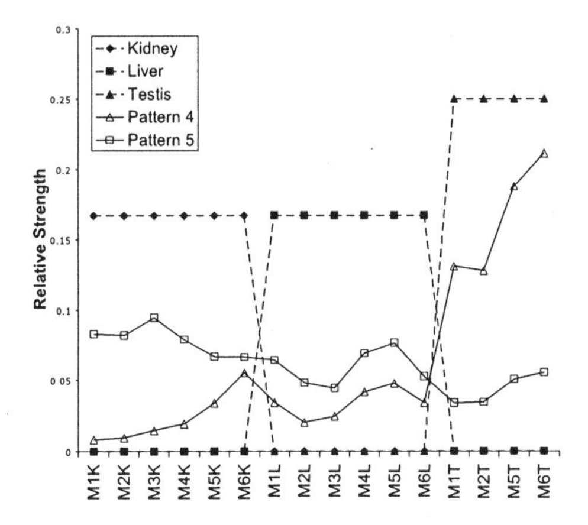

Figure 3. The results of Bayesian Decomposition analysis positing five patterns. The tissue specific patterns are enforced in the analysis. Here they are depicted with dashed lines and solid marks. The two other patterns (Pattern 4 and Pattern 5), depicted with solid lines and open marks, are free to take on any form. All patterns are normalized to a total strength of 1.

3. RESULTS

3.1 Patterns and Distributions

The results for Bayesian Decomposition analysis of the data for five patterns is shown in Figure 3 (the result for six patterns is similar and is discussed below). Pattern 4 (shown with open triangles) represents a pattern related to high expression in the testis, especially in mice five and six. This pattern cannot be explained by the testis pattern because of the significant differences apparent between the first two and last two mice. Under ideal conditions this pattern would not be needed. Pattern 5 is a pattern showing variation between all mice, but with a significant constant level of expression. Interpretation is discussed below.

3.2 Five or Six Patterns

The results for Bayesian Decomposition analysis of the data set with five and six patterns were very similar. Figure 4 shows the results of the three non-tissue specific patterns. The patterns depicted with squares represent the patterns when positing six patterns, while those with triangles represent the results when positing five patterns. The patterns with dotted lines and open markers are virtually identical and represent the pattern related to high expression in the testis, especially in mice five and six.

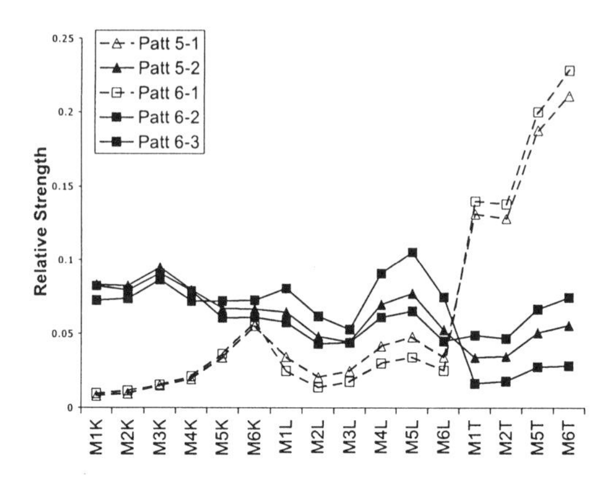

Figure 4. The differences between patterns positing five patterns and positing six patterns. The two patterns related to the differences in mouse testis are shown with open markers and dotted lines. The three patterns with solid lines are very similar (triangle marker from positing five patterns and square markers from positing 6).

The more difficult question involves whether this data set is better modeled by four or by five other patterns. The differences between the three patterns (one from positing five patterns and two from positing six) are not great. In order to determine whether to use five or six patterns, we turned to further analysis using gene ontology information.

3.3 Gene Ontology

Interpreting data in light of gene ontology was done by identifying genes which were present in each pattern. The decomposition of Equation 1 and Figure 1 leads to the patterns depicted in Figures 3 and 4. For each pattern, there is a corresponding column in the distribution matrix that measures the strength of the assignment of each gene to the pattern. Since Bayesian Decomposition relies on Markov chain sampling of the probability space describing the possible **A** and **P** matrices, each strength measurement (value in the **A** matrix) has an associated uncertainty. Here we define a gene to be associated with a pattern if the strength of its assignment to a pattern is three standard deviations above zero.

For all patterns we then determine the number of genes within the pattern assigned to each gene ontology term and compare this to the number of genes assigned to the term within the full data set. In order to avoid major enhancement appearing due to only a few genes being assigned to the gene ontology term, we eliminated all terms represented by less than five genes in the data set (leaving 224 terms out of an original 724 terms). We compute the enhancement of the term within the pattern as

$$\text{Enhancement} = \frac{N_{patt}^{GO} \,/\, N_{patt}^{TOTAL}}{N_{data}^{GO} \,/\, N_{data}^{TOTAL}} \qquad [2]$$

where the numerator gives the normalized number of genes with the GO term in the pattern and the denominator is the normalized number of genes with the GO term in the full data set. We then order the gene ontology terms by the strength of their enhancement and use this to deduce a function or set of functions for the pattern.

Table 1. The non-tissue specific patterns in terms of the enhancement of Gene Ontology terms in order from most enhanced to least enhanced.

Pattern 5-5 (1.3 – 1.4 fold)	protein secretion, secretory pathway, anti-apoptosis, translational initiation, coenzymes and prosthetic group biosynthesis, steroid biosynthesis, N-linked glycosylation, lipid biosynthesis, main pathways of carbohydrate metabolism, catabolic carbohydrate metabolism
Pattern 6-5 (2 – 3.4 fold)	aromatic compound metabolism, main pathways of carbohydrate metabolism, aromatic amino acid family metabolism, catabolic carbohydrate metabolism, energy derivation by oxidation of organic compounds, amino acid catabolism, amine catabolism, coenzymes and prosthetic group metabolism, energy pathways, coenzymes and prosthetic group biosynthesis, micro-tubule-based process, amino acid derivative metabolism, membrane lipid metabolism, fatty acid metabolism, organic acid metabolism, carboxylic acid metabolism, electron transport, glucose catabolism, carbohydrate catabolism, hexose catabolism, alcohol catabolism, mono-saccharide catabolism, protein targeting, protein folding, phospholipid metabolism, regulation of neurotransmitter levels
Pattern 6-6 (2 – 2.7 fold)	Wnt receptor signaling pathway, organic acid biosynthesis, carboxylic acid biosynthesis, oxygen and reactive oxygen species metabolism, polysaccharide metabolism, ecto-derm development, epidermal differentiation, histogenesis, secretory pathway, N-linked glycosylation

The first question to answer was whether to focus analysis on the solution for five patterns or for six patterns. In order to gauge this, we looked at the gene ontology terms enhanced in the non-tissue specific patterns for the five and six pattern sets. Table 1 compares the three patterns in terms of gene ontology. The terms for six patterns are enhanced more sharply as pattern 5-5 splits neatly into two patterns, dropping the number of genes in patterns 6-5 and 6-6 substantially (thus reducing the denominator of the first term in Equation 2). Furthermore the split appears to make some biological sense, as pattern 6-5 appears to be primarily involved in metabolism and catabolism, while pattern 6-6 is more focused on biosynthesis. As such, we chose to focus the analysis on six patterns.

Table 2. Gene ontology terms enhanced in the kidney. For each term the number of genes associated with that term in the data set and the level of enhancement are shown.

Gene Ontology Term	Genes	Fold
amino acid catabolism	7	3.9
amine catabolism	7	3.9
aromatic amino acid family metabolism	6	3.8
aromatic compound metabolism	7	3.2
anion transport	8	2.8
chloride transport	5	2.7
inorganic anion transport	5	2.7
oxygen and reactive oxygen species metabolism	5	2.7
amino acid metabolism	16	2.5
main pathways of carbohydrate metabolism	11	2.5
amino acid biosynthesis	8	2.3
endocytosis	10	2.3
catabolic carbohydrate metabolism	10	2.3
amine metabolism	19	2.1
energy derivation by oxidation of organic compounds	15	2.1
electron transport	15	2.1
amino acid and derivative metabolism	20	2.0
amine biosynthesis	9	2.0

3.4 Tissue Specific Patterns

3.4.1 Enhancement of Gene Ontology Terms

We pursued analysis of the tissue specific patterns in several ways. First, we repeated the analysis of section 3.3 on the patterns enforced to be tissue specific (dashed lines, solid markers in Figure 3). The results are summarized in Tables 2 – 4, where we maintain information on the amount of enhancement for each term and the number of occurrences of the term in the data set.

From the table for the kidney (Table 2), it is clear that transport functions are enhanced within the kidney, as are a number of metabolic functions.

In Table 3, gene ontology terms enhanced within the liver pattern are shown. As with the kidney, there is an enhancement in terms involving amino acid catabolism. In addition, there is enhancement in terms logically linked with the liver, including blood coagulation, hemostasis, and cholesterol and sterol metabolism. In addition, terms related to embryogenesis and morphogenesis are enhanced, possibly related to the fact that the liver is the only organ within the mouse capable of regeneration.

Table 3. Gene ontology terms enhanced in the liver. For each term the number of genes associated with that term in the data set and the level of enhancement are shown.

Gene Ontology Term	Genes	Fold
amino acid catabolism	7	5.5
amine catabolism	7	5.5
aromatic amino acid family metabolism	6	5.3
amino acid metabolism	16	4.8
amino acid biosynthesis	8	4.8
aromatic compound metabolism	7	4.6
amine biosynthesis	9	4.2
amino acid and derivative metabolism	20	4.1
amine metabolism	19	4.0
Lipid transport	7	3.6
amino acid derivative metabolism	8	3.2
blood coagulation	10	3.2
hemostasis	10	3.2
electron transport	15	2.5
biogenic amine metabolism	5	2.5
positive regulation of cell proliferation	11	2.3
embryogenesis and morphogenesis	12	2.1
cholesterol metabolism	6	2.1
sterol metabolism	6	2.1
regulation of neurotransmitter levels	6	2.1
behavior	6	2.1

In Table 4, gene ontology terms enhanced in the testis pattern are shown. The testis is fairly unique among organs in that it has very specialized terms enhanced within the testis pattern, including terms for different cell functions that require constant cell cycle activation and generation of sperm. This specialization is strongly represented within the gene ontology cycle stages, terms for chromatin maintenance, and especially strong enhancement of genes for spermatogenesis, gametogenesis, and reproduction. This reflects the very specialized nature of the testis, both in terms of how it functions and the fact that many genes may not be highly expressed in other tissues.

Table 4 Gene ontology terms enhanced in the testis. For each term the number of genes associated with that term in the data set and the level of enhancement are shown.

Gene Ontology Term	Genes	Fold
spermatogenesis	12	7.7
reproduction	15	6.1
gametogenesis	15	6.1
DNA dependent DNA replication	5	5.2
DNA replication and chromosome cycle	16	4.1
microtubule-based process	13	4.0
S phase of mitotic cell cycle	14	3.7
DNA replication	14	3.7
regulation of cell shape and cell size	19	3.5
nuclear division	12	3.3
cytoskeleton organization and biogenesis	21	3.1
protein kinase cascade	13	3.0
chromatin assembly/disassembly	14	2.8
DNA packaging	19	2.8
nuclear organization and biogenesis	24	2.7
chromosome organization and biogenesis (sensu Eukarya)	24	2.7
organelle organization and biogenesis	29	2.7
cell organization and biogenesis	54	2.7
mitochondrion organization and biogenesis	5	2.6
DNA metabolism	50	2.6
cytoplasm organization and biogenesis	30	2.6
M phase of mitotic cell cycle	10	2.6
mitosis	10	2.6
DNA repair	15	2.6
actin filament-based process	5	2.6
sodium transport	5	2.6
translational initiation	5	2.6
calcium ion transport	5	2.6
mitotic cell cycle	31	2.5
regulation of transcription from Pol II promoter	11	2.4
M phase	22	2.4
establishment and/or maintenance of chromatin architecture	18	2.2
pattern specification	6	2.2
phospholipid metabolism	6	2.2
Wnt receptor signaling pathway	6	2.2
di-, tri-valent inorganic cation transport	6	2.2

3.4.2 Tissue Specific Expression

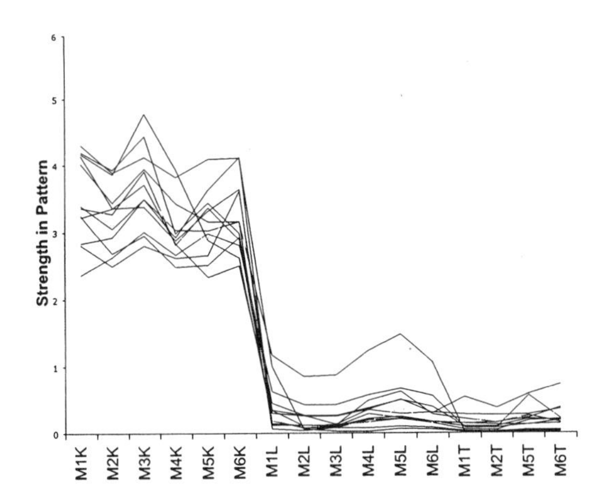

Figure 5. The input data on genes assigned uniquely to the kidney (i.e. >70% of their behavior explained by the kidney pattern alone). The outlier (clone 521796) does show significant expression outside of the kidney (~25% of total behavior), however the 70% cutoff is still exceeded in the kidney.

It is also possible to use Bayesian Decomposition to identify genes that are "uniquely" expressed within a single tissue. Since the decomposition of Equation 1 places all genes into all patterns at some level, this is done by choosing a cutoff. Typically, if a certain percentage of the behavior of a gene is explained by a single pattern, that gene is said to be specific to that pattern. The exact value of the cutoff is dependent on the noise seen within the data. Over the entire data set, we calculated an uncertainty of 27% based on the standard deviation of the mean across replicates. As such, we chose to define a gene as uniquely in a pattern when 70% of its total behavior was explained by a single pattern.

Figure 5 shows the genes uniquely assigned to the kidney pattern. These genes include villin, lipoprotein lipase, alpha-albumin, SA protein, cadherin 16, solute carrier family 12 member 1, folate receptor alpha precursor, insulin-like growth factor 1 receptor, urumodulin, and a few genes of putative structure.

Figure 6 shows genes uniquely assigned to the liver pattern. These genes include alpha-2-macroglobulin, coagulation factor II, retinol binding protein, apolipoprotein CI, sterol carrier protein 2, complement 3 precursor, beta-2 microglobulin, fibrillin 2, and amylase 1.

Figure 7 shows genes uniquely assigned to the testis pattern. These genes include thymosin beta-1, ADP ribosylation-like 4, an EGF-related protein, P4-6 protein, signal transducer and activator of transcription 1, outer dense fiber of sperm tails 1, and protamine 2.

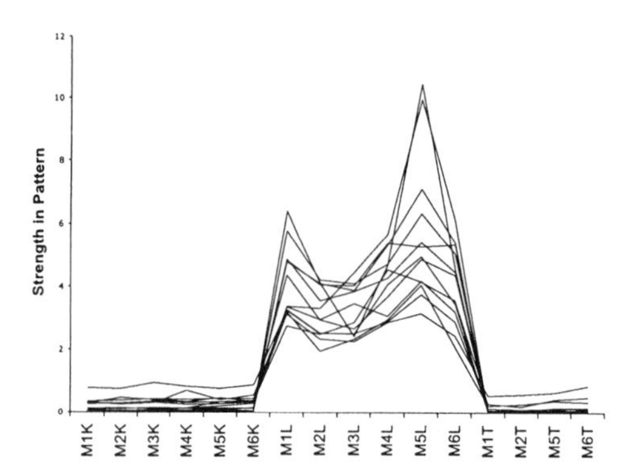

Figure 5. The input data on genes assigned uniquely to the liver (i.e. >70% of their behavior explained by the liver pattern alone).

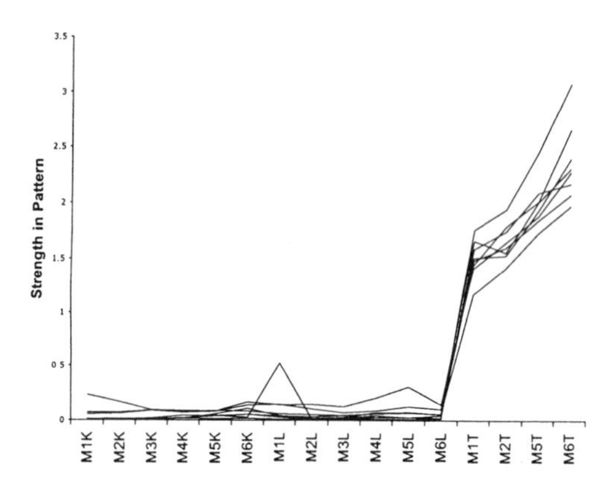

Figure 6. The input data on genes assigned uniquely to the testis (i.e. >70% of their behavior explained by the testis pattern alone).

3.4.3 Using Gene Ontologies

One of the potential pitfalls of using gene ontology information can be demonstrated by looking at the 12 genes that are assigned to the spermatogenesis term in the gene ontologies. Since spermatogenesis is unique to the testis, there is a tendency to think of expression of each gene as logically appearing only in this tissue. However, this is not the case.

Figures 8 and 9 divide these genes into those assigned uniquely by Bayesian Decomposition to the testis and those assigned to multiple patterns.

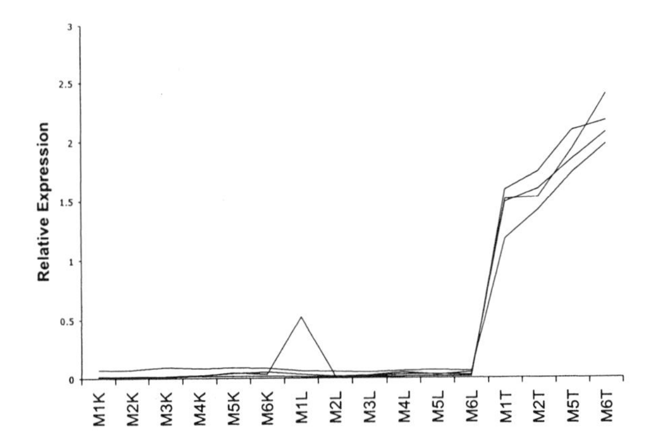

Figure 7. Genes assigned by Gene Ontology to the spermatogenesis term and assigned uniquely to the testis pattern by Bayesian Decomposition. These genes include clones 351811, 602551 (Transition Protein 1), 603127 (Outer Dense Fiber of Sperm Tails 1), and 603132 (Protamine 2).

Figure 9 contains four genes which appear to be unique to the testis tissue in this data set. On closer examination four of these genes are assigned uniquely to the testis pattern and pattern 4, the pattern shown with a dashed line in Figure 4 that appears to be needed to handle excess expression of some genes in testis tissue from mice 5 and 6. These genes then also are unique to the testis in this data set.

The other four genes shown with solid lines in Figure 9 however have expression in multiple tissues. A look at the gene ontologies demonstrate why this occurs. These genes are involved with processes that can be widespread within organisms, including an AMP activated protein kinase (554160), a transmembrane protein involved in proteolysis (571923), a gene involved in oxidative stress response (616656), and a gene encoding a histocompatibility antigen gamma chain (621530).

The reuse by evolution of proteins to perform new functions is commonplace. The advantage of gene ontology therefore arises in the power of looking at many genes simultaneously and observing which terms are enhanced. The appearance of a term is of itself not interesting, unless it appears due to an enhancement of genes containing that term over other possible terms. For instance, in a process involving the genes pictured with solid lines in Figure 9, spermatogenesis will appear as a process. However, it is only when many spermatogenesis genes appear upregulated leading to the enhancement of this term substantially that the process of spermatogenesis is indicated, as it is for the testis tissue.

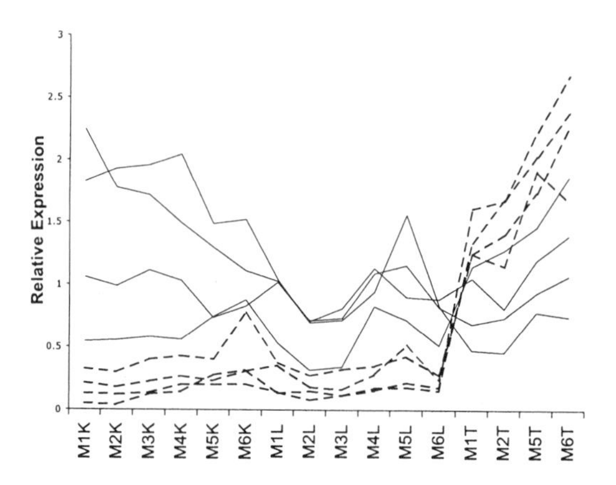

Figure 9. Genes assigned by Gene Ontology to the spermatogenesis term and not assigned uniquely to the testis pattern by Bayesian Decomposition. The genes noted with the dashed line were assigned by Bayesian Decomposition jointly to the testis pattern and to pattern 4, the pattern which handled the oddly high expression in the testis for mice 5 and 6. These four are clones 577547, 582549, 602373 (Testis Specific Protein Kinase 1), and 635148 (Ret Finger Protein).

3.5 Multiple Tissue Expression

Another advantage to the use of Bayesian Decomposition is the ability to remove background patterns (such as those due to routine metabolism and biosynthesis), and focus on the remaining patterns in sets. This allows us to ask about genes which are upregulated in multiple tissues, while ignoring genes involved in background processes (i.e. by removing patterns like those displayed with solid lines in Figure 4 from consideration). Three genes appear strongly expressed in the kidney and liver, but not in testis nor in background processes. These are phenylalanine dehydroxylase, glutaryl coenzyme A dehydrogenase, and a protein similar to cystathionine gamma-lyase. Individual genes can be followed up in the literature. For instance, glutaryl coenzyme A dehydrogenase has been implicated in glutaric acidemia type I and, although ubiquitously expressed, is known to be especially highly expressed in liver and kidney in mice [Woontner *et al.*, 2000].

While no genes appear in common between liver and testis, several appear in common between testis (including the additional pattern) and kidney. They include palmitoyl-protein thioesterase (PPT), a gene highly similar to human profilin II, and metallothionine I. Profilin II in humans is highly homologous to profilin II in mouse. The mouse gene is known to be highly expressed in kidney, in addition to brain, however the testis level has not been reported independently [Lambrechts *et al.*, 2000].

4. DISCUSSION

Exploration of gene expression on a genomic scale is now possible. One outcome of such measurements can be identification of novel genes or novel locations of expression of known genes in an organism. In this case, the expression of the individual gene can be validated by other methods such as RT-PCR. In addition, protein levels can be checked by Western blotting or targeted antibodies when available.

While identification of individual genes showing differential regulation is useful, other possibilities exist for use of the substantial amount of data available in microarray measurements. Recovery of this information generally takes the form of pattern recognition, where identification of complex processes is desired. Bayesian Decomposition is designed to identify underlying processes and assign individual sets of measurements to these processes. Here the sets of measurements are the expression levels of a gene in multiple mice and tissues with the underlying processes being tissue specific and non-tissue specific biological behavior yielding expression of multiple genes. Identification of the patterns requires additional information recovered from biological databases, such as Gene Ontology databases and gene indices. This results from the complexity of the biological data – the measurements allow many possible mathematical interpretations so that the biology must guide the researcher to the correct one.

In section 3.3, the biological knowledge encoded in gene ontologies was used to guide the choice of the number of underlying patterns which best described the data set. Once the number of patterns was chosen, the gene ontologies then permitted interpretation of the patterns related to the specific tissues in section 3.4.1. This validated the patterns as the gene ontology terms enhanced within the individual tissues were those expected given the tissue function.

In addition, Bayesian Decomposition can give results similar to those seen within clustering algorithms as demonstrated in section 3.4.2. By asking for only genes that are unique to a pattern, the researcher can identify specific genes which appear to not be expressed outside the organ of interest (with the caveat that expression was measured in only three organs).

Of greater interest is the ability of Bayesian Decomposition to remove from consideration background behavior. Most expression is actually not related to a process under study (e.g., tumorigenesis) but instead relates to routine cellular behaviors, such as metabolic processes and routine maintenance. These expression patterns can confound analysis, especially as the genes involved in these processes are likely to play a role in the process of interest as well. The ability of Bayesian Decomposition to isolate the background patterns and remove them from further analysis is a key advantage in classification problems. Information such as that in section 3.5 can give indications of how classes are related to one another. For instance, in an analysis of tumor samples, such information could aid in identifying those subclasses of tumors which are most alike as well as indicating which processes they share. Such information can aid in understanding tumorigenesis in specific cases and could in the future aid in decisions on the use of targeted therapeutics.

A key issue in using microarray data to interpret the changes occurring in the underlying biological processes is the need to link expression changes to modifications in signaling or metabolic pathways. Here we have used gene ontology to interpret patterns. However, as noted in section 3.4.3, this can be problematic due to evolution's reuse of the basic material of biological systems. Instead, a more focused set of information will be needed to fully utilize microarrays. This set of information will need to link the expression of specific sets of genes to specific processes and pathways. Bayesian Decomposition can handle separating such sets even in cases where all genes are related to multiple processes if the processes have signatures. The growing databases on pathways can provide some of the information needed, as demonstrated previously in yeast [Bidaut *et al.*, 2002], so that the usefulness of Bayesian Decomposition should continue to improve as other sources of information grow.

ACKNOWLEDGMENTS

We thank the National Institutes of Health, National Cancer Institute (Comprehensive Cancer Center Core Grant CA06927 to R. Young), the State of Pennsylvania (Health Research grant to M. Ochs), and the Pew Foundation for support.

REFERENCES

Alizadeh, AA, Eisen, MB, Davis, RE, Ma, C, Lossos, IS, Rosenwald, A, Boldrick, JC, Sabet, H, Tran, T, Yu, X, Powell, JI, Yang, L, Marti, GE, Moore, T, Hudson, J, Jr., Lu, L, Lewis, DB, Tibshirani, R, Sherlock, G, Chan, WC, Greiner, TC, Weisenburger, DD, Armitage, JO, Warnke, R, Staudt, LM and et al. (2000) Distinct types of diffuse large B-cell lymphoma identified by gene expression profiling. Nature 403: 503-11.

Alter, O, Brown, PO and Botstein, D (2000) Singular value decomposition for genome-wide expression data processing and modeling. Proc Natl Acad Sci U S A 97: 10101-6.

Ashburner, M, Ball, CA, Blake, JA, Botstein, D, Butler, H, Cherry, JM, Davis, AP, Dolinski, K, Dwight, SS, Eppig, JT, Harris, MA, Hill, DP, Issel-Tarver, L, Kasarskis, A, Lewis, S, Matese, JC, Richardson, JE, Ringwald, M, Rubin, GM and Sherlock, G (2000) Gene ontology: tool for the unification of biology. The Gene Ontology Consortium. Nat Genet 25: 25-9.

Besag, J, Green, P, Higdon, D and Mengersen, K (1995) Bayesian computation and stochastic systems. Statistical Science 10: 3 - 66.

Bidaut, G, Moloshok, TD, Grant, JD, Manion, FJ and Ochs, MF (2002). Bayesian Decomposition analysis of gene expression in yeast deletion mutants. In: Methods of Microarray Data Analyis II (Lin SM, Johnson KF, eds), Kluwer Academic Publishers, Boston, 2002: 105-122.

Bittner, M, Meltzer, P, Chen, Y, Jiang, Y, Seftor, E, Hendrix, M, Radmacher, M, Simon, R, Yakhini, Z, Ben-Dor, A, Sampas, N, Dougherty, E, Wang, E, Marincola, F, Gooden, C, Lueders, J, Glatfelter, A, Pollock, P, Carpten, J, Gillanders, E, Leja, D, Dietrich, K, Beaudry, C, Berens, M, Alberts, D and Sondak, V (2000) Molecular classification of cutaneous malignant melanoma by gene expression profiling. Nature 406: 536-40.

Brown, MP, Grundy, WN, Lin, D, Cristianini, N, Sugnet, CW, Furey, TS, Ares, M, Jr. and Haussler, D (2000) Knowledge-based analysis of microarray gene expression data by using support vector machines. Proc Natl Acad Sci U S A 97: 262-7.

Cho, RJ, Campbell, MJ, Winzeler, EA, Steinmetz, L, Conway, A, Wodicka, L, Wolfsberg, TG, Gabrielian, AE, Landsman, D, Lockhart, DJ and Davis, RW (1998) A genome-wide transcriptional analysis of the mitotic cell cycle. Mol Cell 2: 65-73.

Eisen, MB, Spellman, PT, Brown, PO and Botstein, D (1998) Cluster analysis and display of genome-wide expression patterns. Proc Natl Acad Sci U S A 95: 14863-8.

Geman, S and Geman, D (1984) Stochastic relaxation, Gibbs distributions, and the Bayesian restoration of images. IEEE Transactions on Pattern Analysis and Machine Intelligence PAMI-6: 721 - 741.

Getz, G, Levine, E and Domany, E (2000) Coupled two-way clustering analysis of gene microarray data. Proc Natl Acad Sci U S A 97: 12079-84.

Golub, TR, Slonim, DK, Tamayo, P, Huard, C, Gaasenbeek, M, Mesirov, JP, Coller, H, Loh, ML, Downing, JR, Caligiuri, MA, Bloomfield, CD and Lander, ES (1999) Molecular classification of cancer: class discovery and class prediction by gene expression monitoring. Science 286: 531-7.

Hanisch, D, Zien, A, Zimmer, R and Lengauer, T (2002) Co-clustering of biological networks and gene expression data. Bioinformatics 18 Suppl 1: S145-S154.

Heyer, LJ, Kruglyak, S and Yooseph, S (1999) Exploring expression data: identification and analysis of coexpressed genes. Genome Res 9: 1106-15.

Hughes, TR, Marton, MJ, Jones, AR, Roberts, CJ, Stoughton, R, Armour, CD, Bennett, HA, Coffey, E, Dai, H, He, YD, Kidd, MJ, King, AM, Meyer, MR, Slade, D, Lum, PY, Stepaniants, SB, Shoemaker, DD, Gachotte, D, Chakraburtty, K, Simon, J, Bard, M and Friend, SH (2000) Functional discovery via a compendium of expression profiles. Cell 102: 109-26.

Ideker, T, Thorsson, V, Siegel, AF and Hood, LE (2000) Testing for differentially-expressed genes by maximum-likelihood analysis of microarray data. J Comput Biol 7: 805-17.

Kerr, MK, Afshari, CA, Bennett, L, Bushel, P, Martinez, J, Walker, NJ and Churchill, GA (2002) Statistical analysis of a gene expression microarray experiment with replication. Statistica Sinica 12: 203-218.

Kerr, MK and Churchill, GA (2001) Bootstrapping cluster analysis: assessing the reliability of conclusions from microarray experiments. Proc Natl Acad Sci U S A 98: 8961-5.

Kerr, MK and Churchill, GA (2001) Statistical design and the analysis of gene expression microarray data. Genet Res 77: 123-8.

Khan, J, Wei, JS, Ringner, M, Saal, LH, Ladanyi, M, Westermann, F, Berthold, F, Schwab, M, Antonescu, CR, Peterson, C and Meltzer, PS (2001) Classification and diagnostic prediction of cancers using gene expression profiling and artificial neural networks. Nat Med 7: 673-9.

Kossenkov, A, Manion, FJ, Korotkov, E, Moloshok, TD and Ochs, MF (2003) ASAP: automated sequence annotation pipeline for web-based updating of sequence information with a local dynamic database. Bioinformatics 19: 675-676.

Lambrechts, A, Braun, A, Jonckheere, V, Aszodi, A, Lanier, LM, Robbens, J, Van Colen, I, Vandekerckhove, J, Fassler, R and Ampe, C (2000) Profilin II is alternatively spliced, resulting in profilin isoforms that are differentially expressed and have distinct biochemical properties. Mol Cell Biol 20: 8209-19.

Moloshok, TD, Klevecz, RR, Grant, JD, Manion, FJ, Speier, WFt and Ochs, MF (2002) Application of Bayesian decomposition for analysing microarray data. Bioinformatics 18: 566-75.

Park, PJ, Tian, L and Kohane, IS (2002) Linking gene expression data with patient survival times using partial least squares. Bioinformatics 18 Suppl 1: S120-7.

Pritchard, CC, Hsu, L, Delrow, J and Nelson, PS (2001) Project normal: defining normal variance in mouse gene expression. Proc Natl Acad Sci U S A 98: 13266-71.

Sibisi, S and Skilling, J (1997) Prior distributions on measure space. Journal of the Royal Statistical Society, B 59: 217 - 235.

Sivia, DS (1996) Data analysis : a Bayesian tutorial. Oxford, Oxford University Press.

Spellman, PT, Sherlock, G, Zhang, MQ, Iyer, VR, Anders, K, Eisen, MB, Brown, PO, Botstein, D and Futcher, B (1998) Comprehensive identification of cell cycle-regulated genes of the yeast Saccharomyces cerevisiae by microarray hybridization. Mol Biol Cell 9: 3273-97.

Tamayo, P, Slonim, D, Mesirov, J, Zhu, Q, Kitareewan, S, Dmitrovsky, E, Lander, ES and Golub, TR (1999) Interpreting patterns of gene expression with self-organizing maps: methods and application to hematopoietic differentiation. Proc Natl Acad Sci U S A 96: 2907-12.

Woontner, M, Crnic, LS and Koeller, DM (2000) Analysis of the expression of murine glutaryl-CoA dehydrogenase: in vitro and in vivo studies. Mol Genet Metab 69: 116-22.

15

THE USE OF GO TERMS TO UNDERSTAND THE BIOLOGICAL SIGNIFICANCE OF MICROARRAY DIFFERENTIAL GENE EXPRESSION DATA

Ramón Díaz-Uriarte, Fátima Al-Shahrour, and Joaquín Dopazo
Bioinformatics Unit, Centro Nacional de Investigaciones Oncológicas, (CNIO), (Spanish National Cancer Centre), Madrid, Spain; email: {rdiaz, falshahrour,jdopazo}@cnio.es

Abstract: We show one way of using Gene Ontology (GO) to understand the biological relevance of statistical differences in gene expression data from microarray experiments. To illustrate our methodology we use the data from Pritchard *et al.* [2001]. Our approach involves three sequential steps: 1) analyze the data to sort genes according to how much they differ between/among organs using a linear model; 2) divide the genes based on ``how much or how strongly" they differ, separating those more expressed in one organ vs. those more expressed in the other organ; 3) examine the relative frequency of GO terms in the two groups, using Fisher's exact test, with correction for multiple testing, to assess which of the GO terms differ significantly between the groups of genes. We repeat steps 2) and 3) using a sliding window that covers all the sorted genes, so that we successively compare each group of genes against all others.

By using the GO terms, we obtain biological information about the predominant biological processes or molecular functions of the genes that are differentially expressed between organs, making it easier to evaluate the biological relevance of inter-organ differences in the expression of sets of genes. Moreover, when applied to novel situations (e.g., comparing different cancer conditions), this method can provide important hints about the biologically relevant aspects and characteristics of the differences between conditions. Finally, the proposed method is easily applied.

Key words: Gene Ontology, Unigene, DNA microarray, Fisher's exact test, multiple testing, ANOVA, linear models

1. INTRODUCTION

DNA array technology [e.g., Brown & Botstein, 1999] allows us to analyze the behaviour of several thousands of genes simultaneously in a unique experiment. This number of genes constitutes a significant proportion of all the genes being expressed in a tissue and, consequently, gives us the complete picture of the biological processes active in the sample studied. The molecular basis for the different phenotypes (tissues, diseases, organs, etc.), can be attributed to those genes showing a significant differential expression among the phenotypes. This significance is usually obtained by means of a test that provides evidence that the observed differences in behaviour are unlikely to be observed by chance (the *p*-value). Nevertheless some differences in the gene behavior with a clear biological meaning could occur at a level indistinguishable from a random difference and vice versa. If the significance in the distribution of biological processes or functions is used to determine which differences in the expression level are to be considered significant, the test becomes a tool to set a threshold with biological meaning; likewise, if the differential distribution of terms that relate to the biological process or molecular function is used to label sets of genes based on how much they differ (for example, between organs), then we can use these terms to try to understand the biological relevance of differential gene expression.

Gene Ontology provides a powerful source of information to be used to infer differences in gene expression based on biological background. We have used Gene Ontology to investigate the biological relevance of differential gene expression among three mouse organs, and will specifically use the differences between kidney and testis as our main example.

2. GENERAL OUTLINE OF THE METHODOLOGY

Pritchard *et al.* [2001] present microarray gene expression data for three different organs (testis, kidney, liver) of six different mice, with four replicates per organ per mouse. We want to examine what characterizes the genes that differ between the three organs or between pairs of organs (e.g., kidney vs. testis).

2.1 First step: sorting differentially expressed genes

The first step of our method analyzes the microarray experiment data to sort the genes according to how much they differ between organs. In this case, we could use a linear model where we model gene expression as a function of the relevant experimental design features and organ. For instance, we could model

$$y_{ijkl} = \mu + dye_i + mouse_j + organ_k + error_{ijkl} \quad (1)$$

where y_{ijkl} is the $\log_2$(Experimental/Control) ratio that is usual in microarray data analysis, i is the index for dye (i = 1 meaning Cy3 on the experimental channel data, and i = 2 meaning Cy5 on the experimental channel data), j is the index for mouse (i.e., j ={1, 2, 3, 4, 5, 6}), k is the index for organ (e.g., k = 1 for kidney and k = 2 for testis), l is the index for replicate (i.e., l = {1, 2, 3, 4} for each tissue within mouse), μ is a common intercept term, and $error_{ijkl}$ is the random error term. (Some more comments about this model, and its relation to other models, will be discussed below).

The greater the differences between the two organs, the larger the coefficient for organ should be; however, instead of looking directly at the coefficient, we should use the common *t* statistic from a linear regression model which is the coefficient divided by its standard error. This coefficient allows us to compare different genes: genes that are much more expressed in the kidney will have a large positive *t* statistic, and those that are much more expressed in the testis will have a very small (very large in absolute value, but with negative sign) *t* statistic.

In summary, in this first step we sort genes by how much they differ between organs.

2.2 Second step: formation of two groups of genes based on the sorted differences

In the second step we use this sorting information to group genes. We could form one group of genes with all those with $t > 15$ and a second group of genes with all those with $t < -15$. Thus, we would form a first group of genes that are much more expressed in the kidney and a second group that are much more expressed in the testis. Instead of comparing extreme groups (i.e., $t > 15$ against $t < 15$), we might want to compare those genes with $t > 15$ against all other genes; we would therefore be searching for differences between genes that are much more expressed compared to all other genes. The latter is the approach we will use, given that it allows us to highlight what makes a group of genes different (compared to all other genes). In contrast, if we only compare extreme groups (e.g., $t < -15$ vs. $t > 15$), the interpretation of the differences becomes more and more difficult as the threshold (the value of the *t* statistic) becomes smaller.

2.3 Third step: analysis of differential frequency of GO terms

In the third step we examine the relative frequency of Gene Ontology terms between the two groups that we formed in the second step. Gene Ontology (GO), [Ashburner *et al.*, 2000] provides a structured vocabulary for the annotation of genes and proteins. GO terms are structured in a hierarchy (actually, a directed acyclic graph), ranging from more general to more specific. Therefore, in these ontologies a term at a lower level (e.g., level 4) has one (or more) parent term(s) at the preceding level (level 3). GO is structured in three ontologies, corresponding to biochemical function, cellular processes, and cellular components. In Gene Ontology, data can be annotated at varying levels, depending on the information available.

In our study we first selected the GO level for our query. We chose level 3 as it is the best compromise between quantity and quality of GO information [Conde *et al.*, 2002]. We queried the Unigene Id (for mouse genes) for each GO term at that level of resolution and we obtained the GO terms associated with each of our two groups of genes. For each GO term, we tested whether the two groups differ in the frequency of that GO term using Fisher's exact test for 2x2 contingency tables. For each GO term we can represent the data as a 2x2 contingency table with rows being presence/absence of the GO term and each column representing each of the two groups of genes. In other words, the numbers in each cell of the 2x2 contingency table are the number of genes of the first group where the GO term is present, the number of genes of the first group where the GO term is absent, the number of genes of the second group where the GO term is present, and the number of genes of the second group where the GO term is absent.

Fisher's exact test returns the p-value for that contingency table. However, we cannot directly use the individual p-value of each GO term, because we are testing multiple hypotheses, one for each GO term. If we were to use the p-value directly and declare all GO terms with a p-value < 0.05 as significantly differentially represented, we would have a great number of false rejections (i.e., we would end up considering as differentially represented many more GO terms than we should). Thus, we need to account for multiple testing.

There are several methods available to account for multiple testing (reviews in [Dudoit *et al.*, 2002a and b; and Westfall & Young 1993]). We have chosen to control the False Discovery Rate (FDR) using the method of Benjamini & Hochberg [1995]. The FDR is the expected proportion of false rejections (i.e., true null hypothesis that are incorrectly rejected) relative to the total number of rejections. The control of FDR is probably a more appropriate method than other alternatives that control the overall Family Wise Error Rate (the probability that there is one or more false rejections over all the tests conducted), because it is less conservative, and thus it is more appropriate in studies that have a large exploratory component (Benjamini *et al.* 1999). In addition, compared to the resampling-based step-down minP and maxT methods of Westfall & Young [1993]; see also Dudoit *et al.*, [2002a and b], control of FDR using Benjamini & Hochberg's method does not require random permutation of data and is, therefore, much faster computationally. This is an important advantage in a case such as ours, where we repeat the process several hundred times using a sliding window. Using the FDR method, we only considered a GO term as being differentially represented in the two groups of genes if its FDR was smaller than 0.1 (we used the 0.1 level, instead of 0.05 because of the low power of Fisher's exact test and due to the exploratory nature of our study).

In summary, from the third step, we obtain GO terms which have a statistically significantly different relative proportion in the groups of genes.

We can now repeat steps two and three using a different threshold. For instance, if our first division in two groups used thresholds $t > 15$ and $t \leq 15$, we could now form another set of two groups by using instead $t > 10$ and $t \leq 10$. Now we would compare the relative frequency of GO terms between these two new groups. We can repeat the process for a whole set of thresholds. However, since setting up fixed thresholds in the absence of other knowledge might be somewhat arbitrary, we have instead used a sliding window. In this way, by moving over the set of ordered genes, we successively form groups of genes that can be compared to others[1].

In the sections that follow we provide additional details regarding the procedure, using as an example the comparison between kidney and testis. We later comment on alternative approaches that use slightly different steps one and two.

3. DATA PREPROCESSING, STANDARDIZATION, AND REDUCTION OF MISSING VALUES

We obtained the corrected data from Pritchard *et al.* [2001] from the CAMDA'02 web site. In addition, from the authors web site (http://www.pedb.org) we downloaded "Mouse_array_merge_full.txt",

[1] Strictly, we are no longer maintaining control of the family wise error rate, because of the sequence of tests, but we use this as a heuristic device to identify which set of groups have certain terms (e.g., gametogenesis, DNA binding) which are over- or under-represented.

containing the Unigene Ids for each of the clones; we used the Unigene Ids provided by the authors to obtain the GO terms corresponding to each clone.

3.1 Preprocessing and normalization

We first subtracted background from foreground intensity levels for each sample, for both the experimental and control data. Next, using the information about the dye of the experimental channel data, we took the log (base 2) of the ratio experimental/control. Finally, we set all samples with a flag of –50 as missing data. To allow for comparisons among arrays, we standardized the data using the global median for each array (i.e., each of the $\log_2$ ratios were divided by the median $\log_2$ ratio of its array).

3.2 Reduction of missing values

Some analyses of organ effects can be seriously affected by an imbalance in the data sets [Milliken & Johnson, 1992; Miller, 1997], questioning the use of standard F-tests. In addition, and of particular importance for the analyses reported here, the interpretation and use of the t-tests from the regression models is simplified if there is balance in the organ sample sizes; finally, balance in the data makes it easy to compare the differences between pairs of organs.

Our main concern here is the imbalance in the representation of each organ. Thus, we have filtered the data to only use genes where the organ with the smallest sample size has a sample size that is at least 87.5% of the sample size of the organ with the largest sample size (in the case of a gene where at least one of the organs has no missing data, the organ with the largest number of missing data would be allowed to have at most three missing values: 21/24). After applying this filter, we are left with a total of 4140 genes. (This simple criterion additionally ensures that, of these, 3998 genes have at most a total of six missing values.) More restrictive criteria could be used (e.g., eliminating all those genes with a total of more than six missing values or filtering also with respect to mouse), but we have tried to achieve a balance between reliability and interpretability of results, ease of implementation of the procedure, and keeping most of the genes in the data set. Finally, after excluding those genes without Unigene Id information, we were left with 3736 clones for further analyses.

4. FITTING THE LINEAR REGRESSION MODEL

The model we fitted to kidney and testis data was discussed in section 2.1:

$$y_{ijkl} = \mu + dye_i + mouse_j + organ_k + error_{ijkl}. \quad (2)$$

In contrast to Pritchard *et al.* [2001], the dye term was fitted independently to each gene. This dye term can be included in the model because dye swapping was used in the experiment. We choose not to include interactions in this model to facilitate the interpretation of the coefficients. We parameterized the above model so that a large positive coefficient (and thus t statistic) for `organ´ means that the gene is much more expressed in the kidney and a large negative coefficient for organ means that the gene is much more expressed in the testis.

Some of the terms in our model are equivalent to those in the model by Kerr & Churchill [2001 a, b], where they use ANOVAs for the analysis of microarray data: in our analyses, the `dye´ term is equivalent to their DG interaction term, the `mouse´ term is equivalent to the AG interaction term, and the `organ´ term is equivalent to the VG interaction term. As in the models of Kerr & Churchill, we use the `dye´ and `mouse´ term to control for systematic effects that would, otherwise, be assigned to the error term (thus decreasing statistical power), but the `dye´ and `mouse´ terms are of no intrinsic interest. Our interest lies in the organ term, which is the one that measures the strength and direction of testis vs. kidney effect. Instead of the model above, we could have applied the mixed-effects model of Wolfinger *et al.* [2001], and we would then have used their GT term (where their T, for treatment, would have corresponded to our tissues).

We applied the model above to each of the 3736 genes and we divided genes in groups by choosing a threshold. We can interpret the meaning of this threshold examining the magnitude and sign of the t value. Large and positive values correspond to genes that are much more expressed in the kidneys and large and negative values to genes that are much more expressed in the testis. We used a sliding window of 150 genes, which was moved in jumps of 10 genes over the list of all genes ordered from minimum (largest and negative) t value to maximum t value. Given that the windows are of 150 genes and we move in jumps of 10 genes, there is an overlap of 140 genes between successive windows. A window of 150 genes was chosen as a compromise between low power of the Fisher test (see below and discussion) and too large a group that would become too heterogeneous. The number of genes with GO terms in this data was around 20%, so choosing a window of 150 results in about 30 genes with GO in the given window (and about 700 genes with GO in the reference group). A smaller window might result in too few genes with GO in the window group and thus very low power to detect differences in the frequency of terms; on the other hand, too large a group might result in groups of genes that could be too heterogeneous to show a common pattern with respect to biological meaning. Similar qualitative results were obtained with a sliding window of 300 genes.

We might have preferred to use (unadjusted) *p*-values to order genes instead of the *t* statistic, because there are minor differences in the sample sizes, and thus degrees of freedom. We used the *t* statistic directly, however, because these differences in sample size are minor (see above), and thus the relation *t* statistic--*p*-value is essentially the same for most *t* statistics; in this case, using the *t* statistic results in a much simpler procedure with a more straightforward interpretation than using the *p*-values taking into account the sign of the *t* statistic. Moreover, we are using the threshold as a device to form groups, but the ultimate judgment concerning the relevance of the groups is derived from the results with the GO terms.

4.1 Difference in the frequency of GO terms in each group

For each sliding window we compared the frequency of GO terms of the genes included in the window vs. the frequency of GO terms in the genes outside the window. In other words, we examined if there were significant differences (adjusted for multiple testing) in the representation of GO terms in the two groups. We only considered those GO terms that had an FDR-adjusted p-value of less than 0.1[2] as differentially represented. We used the 0.1 level, instead of 0.05 because of the low power of Fisher's exact test and due to the exploratory nature of our study. In addition, we only show those terms that had an adjusted p-value of less than 0.1 in at least three windows, to try to minimize spurious significant results. The results are shown in Figure 1.

The results from molecular function at level 4 are straightforward: the terms "globin", "oxidoreductase, CH-OH group", and "oxygen transporter" are significantly more common, with a difference of about 10%, in those genes that are much more strongly expressed in the kidney. The results for molecular function at level 3 are much richer; genes that are much more expressed in the kidney show more abundance of the "oxygen binding" and "oxydoreductase" terms but show a smaller frequency of the "hydrolase"

[2] Similar qualitative results were obtained using 0.05 instead of 0.1 as the adjusted p-value cut-off.

and "nucleic acid binding" terms. Genes that are more strongly expressed in the kidney also have a somewhat higher frequency of the term "blood coagulation factor." Genes more strongly expressed in the testis show higher prevalence of the terms "structural constituent of ribosome," "structural constituent of cytoskeleton" and "ribonucleoprotein." Finally, genes that are not expressed differentially between the kidney and testis show a higher frequency of the GO term "neurotransmitter binding," which might be a case of a false positive. Most of these results are in agreement with our biological knowledge of the differences between these two organs.

Regarding the biological process, the results for level 3 indicate that the genes that are much more expressed in the kidney have a higher prevalence of the term "transport" but a smaller frequency of the term "metabolism," whereas those much more expressed in the testis have a higher frequency of the term "reproduction." It is somewhat surprising, however, that genes that are not differentially expressed show a lower frequency of the term "metabolism." Level 4 unravels a more complex picture. Genes much more expressed in the testis show a higher frequency of the term "gametogenesis" (whose parent term at level 3 is "reproduction"), and genes more expressed in the testis, but not so extremely, show a higher frequency of the term "M phase" and "muscle contraction." Genes much more expressed in the kidney show a higher frequency of "gas transport" (whose parent term at level 3 is "transport") and a lower frequency of "nucleobase, nucleoside, ..." (whose parent term is "metabolism"). Genes moderately more expressed in the kidney are enriched in the term "phosphate metabolism," and those genes only slightly more expressed in the kidney have a higher frequency of the term "pattern development."

Again, these results agree with our biological knowledge. Notice how the genes that show the highest expression in the kidney compared to the testis are those which have an under-representation of GO terms related to the processes involved in the formation of spermatozoids, but show an overrepresentation of terms involved in transport. Conversely, genes that show the highest expression in the testis have an over-representation of terms related to reproduction and cell cycle (reproduction, gametogenesis, M phase.) Interestingly, other terms (pattern specification and phosphate metabolism) seem to be over-represented in genes that are more (but not extremely more) expressed in the kidney than in the testis. Metabolism being under-represented in genes that show almost no difference between the two organs might be either a false positive, or an artifact due to these sets of genes being enriched in genes with very low expression in most conditions, and thus unlikely to be involved in metabolic processes (which are common to all cells.)

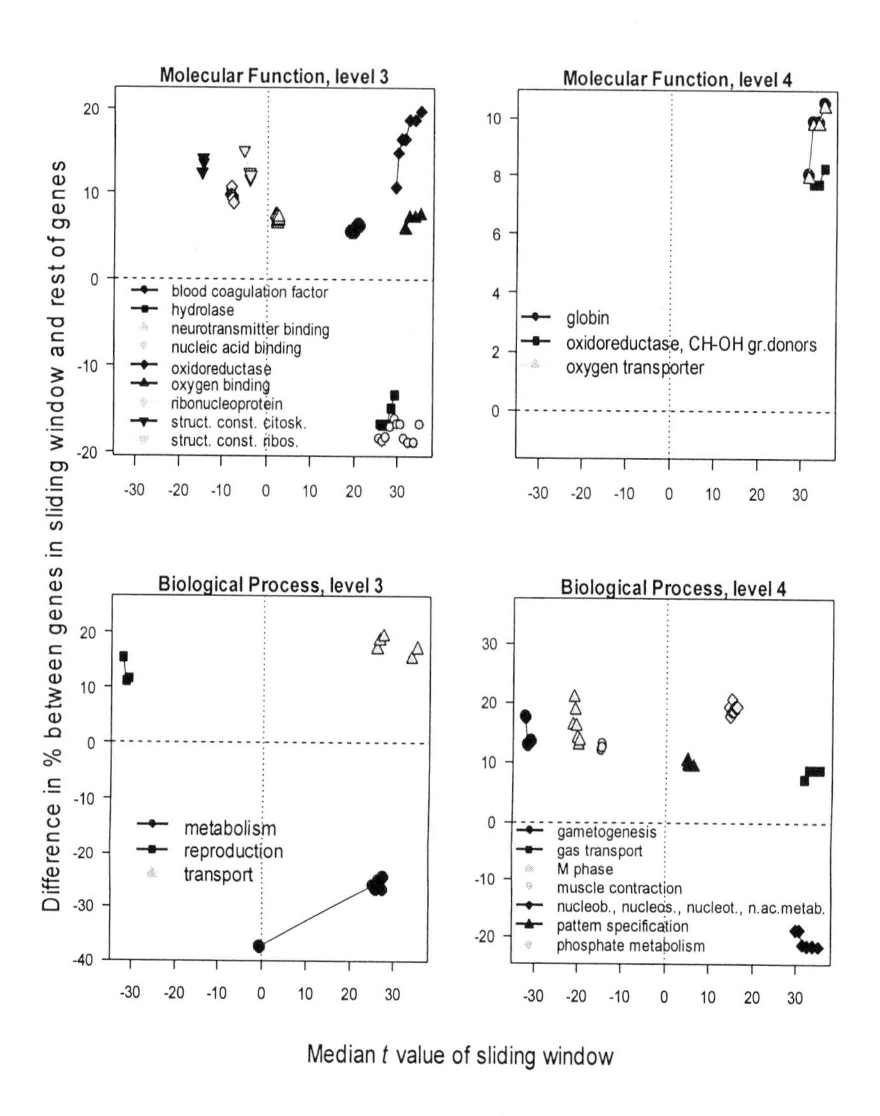

Figure 1. Results for all GO terms that had at least three adjusted p-values < 0.1. Large negative t values correspond to genes much more expressed in the testis, and large positive t values to genes much more expressed in the kidney.

5. ALTERNATIVES TO STEPS ONE AND TWO

5.1 An alternative way of comparing two organs

Instead of using the signed t statistic, we could form groups based upon the p-value. We would use a sliding window over the genes ordered by increasing p-value, so that as we move over the window we would progressively examine genes that are less differentially expressed between organs, regardless of whether they are more or less expressed in one organ or the other. Here we would be comparing differentially expressed genes against the reference group of all other genes, whereas with the linear regression approach we compared genes that are more expressed, with similar strength, in one organ vs. the rest of the genes.

Related to this approach, instead of using the t statistic from the model in equation (1) we could use the F-ratio for organ; genes with a large F-ratio for organ will be those where there are large differences in expression among organs. Of course, using the p-value from the linear regression model above would be equivalent to fitting an ANOVA model and using the p-value for organ.

The problem with these approaches is that, when comparing two groups, the F ratio and the p value do not provide information on directionality of the difference ---i.e., whether it is the testis or the kidney that show larger expression. However, most biologists will generally be interested in the directionality of the difference because it allows for much easier interpretation; thus the use of the t statistic will generally be more appropriate than the use of the F ratio or the p value.

Another alternative is not to use a sliding window, but to compare successive pairs of extreme groups of genes as defined by a threshold; for instance, we could compare genes with a t value < -20 vs. genes with a t value > 20 or, in other words, genes that are much more expressed in the kidney vs. genes that are much more expressed in the testis. However, we suspect that using the sliding window approach we have employed here might better highlight specific biological aspects of groups of genes that differ, in a similar way, between the two organs.

5.2 Comparing three organs

We could fit an ANOVA model for the three organs simultaneously, and using a sliding window over the ordered F values, order genes based on how strongly they show differential expression among the three organs. This is similar to the use of p-values in the previous section, except interpretation might be more complicated because, in addition to losing information on directionality (in which organ is expression the greatest?), the comparison is among three organs.

5.3 Are there multiple testing issues when sorting genes?

Multiple testing issues are not very relevant in the second step, since the ultimate arbiter of the biological significance is the analysis of GO terms. Threshold values for the statistics could be chosen so that they are significant even when considering multiple testing, but genes that are statistically significantly different are not necessarily biologically significant or relevant.

6. DISCUSSION: MERITS, LIMITATIONS, CONCLUSION

The suggested procedure is easy to use. The first two steps can be easily implemented, for instance, using a statistical package with programming facilities; we have used R, but other statistical packages are probably also suited. The third step might seem the most complicated, but we have built a tool, FatiGO [Al-Shahrour *et al.*, in prep.], now available at http://fatigo.bioinfo.cnio.es, that carries this out (searching for GO terms from a list of gene id's ---Unigene Id or Gene Symbol---, and performing Fisher's exact test plus the permutation test for step-down minP multiple testing).

There are two main limitations of the method. The first one is related to the amount and quality of annotation. On the one hand, many genes have no GO annotation; for instance, at threshold 30, only 25 of the 88 genes from the kidney group have an annotation at level 4. In these examples, about 16% to 25% of the genes are annotated at level 4, and about 20% to 30% are annotated at level 3. Of course, the validity of the method with non-annotated genes relies on the non-annotation being independent of the GO term (i.e., that a gene not having a GO annotation be due to reasons unrelated to the value that the GO term would have had). One of the effects of this lack in data is that it makes it hard to detect GO terms that have different frequencies between two groups (an effect that is probably more serious when the original groups of genes are already small ---e.g., with a threshold of 30 in our case). This can also explain some inconsistencies and the fact that some terms might be significant at a given threshold, not significant at the next, and significant again at the following threshold. Finally, this limitation, as it affects smaller groups more strongly, might have its most adverse effects when examining the most extreme differences in expression, which might be those that are of more interest to us. On the other hand, the quality of annotation varies: some genes are annotated manually and carefully, whereas other genes are annotated automatically and based on similarities with other genes, which can lead to dubious annotations (Gene Ontology includes information related to the type of annotation that allows ranking of quality of annotation). We expect some of these limitations to be of lesser importance with time as more genes are annotated, and the annotation of other genes is reviewed. In the present analyses, we have ignored the issues of lack of annotation and of dubious annotations. By using a Bayesian approach, however, it should be possible to explicitly deal with (incorporate) the lack of annotation and annotations of different quality. We are currently working on this.

The second limitation was previously mentioned: since we repeat the process of examining the significance of GO terms many times (one for each sliding window), we are no longer maintaining strict control of the false discovery rate. Providing strict control of the FDR over all the comparisons is probably unwarranted. We suggest that our methodology be used as a heuristic device to identify for what set of groups certain terms (e.g., gametogenesis, DNA binding) are over- or under-represented (providing control of the error rate at each of the threshold levels.) The patterns identified should probably show consistency across adjacent threshold levels and between gene ontology levels, whereas spurious results (due to chance from multiple testing) are likely to show up as occasional significant results that are not consistent across thresholds or gene ontology levels. Given the exploratory nature of the method, and the loss of power associated with the relative scarcity of GO annotations, we prefer to decrease Type II errors (missing a real biological difference) even if that means slightly increasing Type I error rates (declaring significant a term that is not really differentially represented); this way, we provide an increased ability to understand the biological meaning of the detected differences, while limiting (via control of FDR at each window) the number of false detections.

6.1 Similar approaches

Our approach is similar to many other recent proposals that are trying to incorporate annotation data to gene expression data [e.g., Zhou *et al.*, 2002; Gibbons and Roth, 2002; etc]. The most similar was that suggested by Pavlidis *et al.* [2002]. These authors first use a statistical model (in their case an ANOVA) to obtain the *p*-values for each gene. While Pavlidis *et al.*

[2002] compared tumor types and brain regions, we could compare pairs of organs, or the three organs together with an ANOVA. They compute the score for each "GO class," where a GO class is the group of genes that share a GO term; the score for a GO class is the average of the minus $\log_{10}$ p-value of the genes in the GO class, so that GO classes with a large score are composed of genes with lower *p*-values. Finally, Pavlidis *et al.* select classes using those with a relevant score based on a permutation test. Their approach differs from ours on several counts. First, since the p-value is based on an ANOVA, directionality information is lost, so a given GO class can be composed of some genes that are much more expressed in one organ and some other genes that are much more expressed in other organs. This, however, could be solved using the t statistic when comparing two classes, and requiring GO classes to have only t statistics with the same sign. Second, and more importantly, their approach is not designed to answer the general question "what GO terms differentiate this set of genes from that set of genes," but the question "which GO terms are associated with genes that, on average, show strong differential expression." Note that the approach of Pavlidis *et al.* might detect a GO class that includes some genes that show no differential expression between conditions (if the other genes in the GO class do show strong effects). Thus, their approach is a different, and complementary, way of incorporating GO information to gene expression data.

To conclude, we think that our proposed method is promising in identifying biological features of genes that are differentially expressed between/among organs or conditions, since it incorporates information on the known (annotated) biological processes and molecular functions associated with the genes that belong to different groups.

ACKNOWLEDGMENTS

A. Dopazo and L. Lombardía answered questions about GENEPIX. We also thank assistants at CAMDA '02 for their thought-provoking questions, and the editors and anonymous reviewers for comments that improved the manuscript. A. Wren revised the English. R.D-U is contracted by the Ramón y Cajal programme from the MCYT. F. A. is supported by contract BIO2001-0068 from the MCYT.

REFERENCES

Ashburner, M., Ball, C., Blake, J., Botstein, D., Butler, H., Cherry, J., Davis, A., Dolinski, K., Dwight, S., Eppig, J., Harris, M., Hill, D., Issel-Traver, L., Kasarskis, A., Lewis, S., Matese, J., Richardson, J., Ringwald, M., and Rubin, G. S. G. 2000. Gene ontology: tool for the unification of biology gene ontology: tool for the unification of biology. *Nat. Genet*, 25:25--29.

Benjamini, Y. and Hochberg, Y. 1995. Controlling the false discovery rate: a practical and powerful approach to multiple testing. J. Royal Statistical Society, Series B, 57: 289-300.

Benjamini, Y., Drai, D., Kafkafi, N., Elmer, G., Golani, I. 1999. Controlling the false discovery rate in behavior genetics research. (available from http://www.math.tau.ac.il/~ybenja).

Brown, P., O., and Botstein, D. 1999. Exploring the new world of the genome with DNA microarrays. *Nature Biotechnol*, 14:1675--1680.

Conde, L., Mateos, Á., Herrero, J. & Dopazo, J. 2002 Unsupervised reduction of the dimensionality followed by supervised learning with a perceptron improves the classification of conditions in DNA microarray gene expression data. *Neural Networks for Signal Processing XII.* IEEE Press (New York). *Eds. Boulard, Adali, Bengio, Larsen, Douglas.* pp. 77-86

Dudoit, S., Shaffer, J. P., Boldrick, J. C. 2002 Multiple hypothesis testing in microarray experiments. Technical report # 110. Division of Biostatistics, UC Berkeley.

Dudoit, S., Yang, Y. H., Callow, M. J., and Speed. T. P. 2002 Statistical methods for identifying differentially expressed genes in replicated cDNA microarray experiments. Statistica Sinica, 12: 111-139.

Gibbons, F. D., Roth, F. P. 2002. Judging the quality of gene expression-based clustering methods using gene annotation. Genome Research, 12: 1574-1581.

Kerr, M. K., and Churchill, G. A. 2001 a. Experimental design for gene expression microarrays. Biostatistics, 2: 183-201.

Kerr, M. K. and Chucrchill, G. A. 2001 b. Statistical design and the analysis of gene expression microarray data. Genetical Research, 77: 123-128.

Miller, R. G.. 1997. *Beyond Anova* Chapman & Hall.

Milliken, G. A. and Johnson, D. E. 1992. *Analysis of Messy Data*. Chapman & Hall.

Pavlidis, P., Lewis, D. P., Noble, W. S. 2002. Exploring gene expression data with class scores. Proc. Pacific Symp. Biocomputing, 2002: 474-485.

Pritchard, C. C., Hsu, L., Nelson, P. S. 2001. Project normal: defining normal variance in mouse gene expression. *PNAS*, 98:13266—13271.

Westfall, P. H. and Young, S. S. 1993. *Resampling-based multiple testing: examples and methods for p-value adjustment*. John Wiley & Sons.

Wolfinger, R. D., Gibson, G., Wolfinger, E., Bennett, L., Hamadeh, H., Bushel, P., Afshari, C., Paules, R. S. 2001. Assessing gene significance from cDNA microarray expression data via mixed models. Journal of Computational Biology, 8: 625-637.

Zhou, X., Kao, M.-C. J., Wong, W. H. 2002. Transitive functional annotation by shortest-path analysis of gene expression data. PNAS, 99: 12783-12788.

Acknowledgments

The editors would like to thank the contributing authors for their outstanding work. We would also like to acknowledge Emily Allred for her unflagging diligence and assistance in bringing this volume together, as well as for her hard work and dedication in organizing the entire CAMDA conference. We would also like to thank our supporters at Duke University; Darell Bigner, Director Pro Tempore of the Duke Comprehensive Cancer Center and Jim Siedow at the Center for Bioinformatics and Computational Biology. The CAMDA conference would not be possible without the contributions of the scientific committee and other reviewers (listed below) who contribute to the scientific review process. Our thanks for the time and effort they commit to CAMDA. We especially thank our corporate sponsors, Affymetrix, Agilent Technologies, and Glaxo Smith Kline for their generous support. We gratefully acknowledge the North Carolina Biotechnology Center for providing a generous meeting grant.

Reviewers

Bruce Aranow (U Cincinatti)
George Bobashev (RTI)
Philippe Broet (INSERM)
Kevin Coombes (MDACC)
Chris Corton (CIIT)
Joaquin Dopazo (CNIO)
J. Gormley (MBI)
Greg Grant (U Penn)
Susan Halabi (Duke)
Patrick Hurban (Paradigm Genetics)
Stuart Hwang (COR Therapeutics)
Ed Iversen (Duke)
Bret Jessee (AnVil Informatics)
Warren Jones (UAB)
Wendell Jones (Expression Analysis)
Elana Kleymenova (CIIT)
D. P. Kreil (EBI)
Michael Ochs (Fox Chase Cancer Center)
Richard (Rick) Paules (NIEHS)
Raymond Samaha (Applied Biosystems)
Jennifer Shoemaker (Duke)
Paul Wang (Duke)
Dawn Wilkins (U Mississippi)
Thomas Wu (Genentech)
Fei Zou (UNC).

Index